Multiple Choice Questions on Chemistry

1000 Questions and Answers for Entrance Exams

Kingsley Augustine

INTRODUCTION

The chemistry questions in this book have been carefully selected to help students prepare for school examinations and entrance exams into higher institutions of learning. The answers to the questions are provided at the end of the book.

TABLE OF CONTENT

QUESTIONS

1) Equilibrium is said to be attained in reversible reaction when   A. all the reactants have been used up   B. all the products have been formed   C. there is no further change in temperature   D. the rates of the forward and backward reactions are equal   E. the rate of formation of the products decreases with time.

2) An acid present in proteins is   A. lactic acid   B. amino acid   C. propanoic acid   D. palmitic acid   E. stearic acid.

3) The following are uses of sulphur except   A. manufacture of tetraoxosulphate (VI) acid   B. prevention of the growth of fungi   C. Vulcanization rubber   D. manufacture of dyes   E. Coating of steel to prevent rusting.

4) The reaction represented by the equation, $C_6H_{12}O_6 -----> 2C_2H_5OH + 2CO_2$ in the presence of zymase at 25^0C is known as   A. hydrolysis   B. oxidation   C. reduction   D. condensation   E. fermentation

5) The reaction represented by the equation: $NaOH_{(aq)} + HCl_{(aq)} ---> NaCl_{(aq)} + H_2O_{(l)}$ A. is a double decomposition   B. is a neutralization   C. is reversible   D. is usually catalysed   E. attains equilibrium after a few seconds.

6) Which of the following compounds is aromatic?   A.   Benzene   B.   Cyclobutane   C. Cyclopentane   D. Hexane   E. Ethene

7) Which of the following does not affect the rate of a chemical reaction?   A. Concentration of the reactants   B. Addition or presence of a catalyst   C. Size of the reacting particles   D. The enthalpy change for the reaction   E. Temperature at which the reaction is carried out

8) Given that 32.0g sulphur contains 6.02×10^{23} sulphur atoms, how many atoms are there in 2.70g of aluminum? [Al = 27, S = 32]   A. 6.02×10^{23} B. 3.01×10^{23} C. 6.02×10^{22} D. 5.08×10^{22} E. 3.01×10^{22}

9) The empirical formula of a hydrocarbon containing 0.12 moles of carbon and 0.36 moles of hydrogen is   A. CH_2 B. CH_3 C. C_2H_4 D. C_2H_4 E. C_2H_6

10) A salt which loses mass when exposed to air is   A. hygroscopic   B. deliquescent   C. efflorescent   D. fluorescent   E. effervescent

11) Detergents are better than soaps for laundry because   A. detergents are synthetic while soaps are not   B. detergents are more soluble in water than soaps   C. scum is precipitated when soaps are used with hard water but not with detergents   D. soaps are in bars while detergents are in powder form

4

12) Which of the following substances is used as an anaesthetic?   A. CHI_3 B. $CHCl_3$ C. $CHBr_3$ D. CH_3OH E. C_2H_5OH

13) Which of the following types of hybridization gives rise to tetrahedral molecules? A. sp   B. sp^2 C. sp^3 D. sp^3d E. sp^3d^2

14) Which of the following atoms has no neutron?   A. Zinc   B. Neon   C. Sulphur   D. Hydrogen E. Oxygen

15) Which of the following is a neutral oxide?   A. Nitrogen (IV) oxide B. Carbon (IV) oxide   C. Sulphur (VI) oxide D. Phosphorus (V) oxide   E. Nitrogen oxide

16) Which of the following statements is not correct?   A. Carbon exhibits allotropy B. Sulphur exhibits allotropy   C. Chlorine exhibits allotropy D. Nitrogen is a gas E. Argon is a noble gas

17) Which of the following is not true of the elements represented by the symbols
$_b^aX$ and $_b^cY$?
A. X and Y have the same number of electrons B. X and Y have the same number of protons C. X and Y are isotopes   D. X and Y represent the same element E. X and Y represent atoms of different elements

18) The following are uses of radioactive isotopes except for   A.   sterilization   of   medical equipment   B. tracing reaction paths   C. dating elements D. radiography E. determining equilibrium positions

19) The main characteristic feature of transition metals is that they   A. have the same atomic size   B. are reducing agents   C. form ions easily   D. have variable oxidation states   E. are highly electropositive

20) The following are heavy chemicals except   A. tetraoxosulphate (VI) acid   B. caustic soda C. sodium trioxocarbonate (IV) D. ethene E. dyes

21) What quantity of silver is deposited when 96500C of electricity is passed through a solution containing silver ions?   (Ag = 108, IF = 96500C)   A. 1.08g B. 5.40g C. 10.8g D. 21.6g E. 108g

22) Bubbling of carbon (IV) oxide into calcium hydroxide solution results in the formation of A. $CaCO_3$ B. $CuCO_3$ C. $ZnCO_3$ D. Na_2CO_3 E. $MgCO_3$

23) Which of the following is not decomposed by heat?
A. $CaCO_3$ B. $CuCO_3$ C. $ZnCO_3$ D. Na_2CO_3 E. $MgCO_3$

24) What volume of oxygen will be left unreacted when a mixture of $100cm^3$ of hydrogen and $200cm^3$ of oxygen are exploded in a eudiometer?   A. $300cm^3$ B. $200cm^3$ C. $150cm^3$ D. $100cm^3$ E. $50cm^3$

25) Water gas is a mixture of   A. CO and H_2 B. CO and H_2O C. CO_2 and CO D. CO and N_2

26) How many isomers can be obtained from C_4H_{10}?
A. 0   B. 1   C. 2.   D. 3   E. 4

27) Which of the following metals cannot be extracted from its ore by using carbon (II) oxide as the reducing agent?   A. Cu   B. Zn   C. Al   D. Pb   E. Fe

28) Aqueous sodium trioxocarbonate (IV) solution is alkaline because   A. the acid from which it is formed is strong   B. sodium is an alkali metal   C. sodium trioxocarbonate (IV) is decomposed by water   D. sodium trioxocarbonate (IV) is stable to heat   E. sodium trioxocarbonate (IV) is hydrolysed in water

29) Which of the following correctly explains why iodine crystals change directly into the gaseous state when heated?   A. iodine crystals sublime   B. iodine atoms are held together by strong forces   C. iodine molecules are held together by weak forces   D. iodine molecules are covalent   E. iodine is a solid at room temperature

30) Which of the following is a waste product in the Solvay Process for the manufacture of sodium trioxocarbonate(IV)?   A. Ammonium chloride   B. Brine   C. Limestone   D. Calcium oxide   E. Calcium chloride

31) The complex salt formed when aluminium dissolves in sodium hydroxide solution is   A. $Na_3Al(OH)_4$ B. $Na_2Al(OH)_3$ C. $NaAl(OH)_3$ D. $Na_4Al(OH)_4$ E. $NaAl(OH)_4$

32) During the electrolysis of dilute tetraoxosulphate(VI) acid solution 0.05 mole of electrons were passed. What is volume of oxygen released at s.t.p? (Molar volume of a gas at s.t.p. = $22.4dm^3$)   A. $0.224dm^3$ B. $0.280dm^3$ C. $0.560dm$ D. $2.24dm^3$ E. $22.4dm^3$

33) The major raw material in a plastic industry is   A. ethanol   B. sulphur   C. methylethanoate   D. ethene   E. ethane

34) Which of the following statements best explains cracking?   It is a process
A. in which an alkene adds on itself repeatedly to form long chain polymers   B. in which large hydrocarbon molecules are broken into smaller units   C. for measuring the octane number of petrol   D. in which carbon chain results in a zigzag arrangement   E. in which aromatic hydrocarbons are produced

35) Which of the following statements is not correct?   A. Metals high up in the redox series are usually extracted by electrolysis   B. Lead is less reactive than magnesium therefore it is

6

lower in the redox series   C. Aluminium reduces iron (III) oxide to iron in the thermit process. D. An amphoteric oxide exhibits both acidic and basic properties   E. Hydrogen is more reactive than copper therefore it is higher than copper in the electrochemical series.

36) Which of the following hydrocarbons is not likely to be present in petrol?
A. $C_{14}H_{30}$ B. $C_{10}H_{22}$ C. C_9H_{20} D. C_8H_{16} E. C_5H_{12}

37) The most suitable method to use when separating an insoluble solid from a liquid is   A. evaporation   B. filtration   C. magnetization   D. sublimation   E. distillation

38) The following are miscible with water except   A. ethylethanoate   B. methanol   C. ethanoic acid   D. methanoic acid   E. ethanol

39) If $200cm^3$ of a gas at s.t.p. has a mass of 0.268g, what is its molar mass? (Molar volume of a gas at s.t.p. = $22.4dm^3$)   A. 300g   B. 200g   C. 150g   D. 30g   E. 15g

40) How many carbon atoms are there in a benzene ring?
A. 4   B. 5   C. 6   D. 7   E. 8

41) Which of the following gases will bleach moist blue litmus paper?
A. Cl_2 B. CO_2 C. SO_3 D. HCI   E. N_2O

42) When an element can exist in two or more forms in the same physical state, the element is said to exhibit   A. isotopy   B. isomerism   C. allotropy
D. hygroscopy   E. sublimation

43) Which of the following carbohydrates does NOT usually occur in the crystalline form?
A. Frutcose   B. Glucose   C. Surcrose   D. Maltose   E. Cellulose

44) Copper can best be purified by   A.   roasting the impure copper in a blast furnace   B. heating the oxide with coke   C. electrolyzing a copper (II) salt solution using the impure copper as the anode   D. converting the impure copper to a trioxonitrate (V)   E. washing the impure copper with sodium hydroxide solution

45) Locally produced soap in which wood ash is used as the base is usually soft because the wood ash contains a lot of   A. sodium ions   B. potassium ions   C. lithium ions   D. calcium ions   E. magnesium ions

46) Why is potassium fluoride added to the electrolyte in the extraction of sodium from fused sodium chloride?   A.   pure sodium chloride does not conduct electricity   B. Potassium fluoride prevents the sodium produced from attacking the electrodes   C. Potassium fluoride increases conductivity   D.   potassium fluoride reduces the temperature at which sodium chloride fuses   E. the potassium fluoride acts as a catalyst

7

47) The rate of the production of hydrogen in the reaction $Zn_{(s)} + 2HCl_{(aq)} \longrightarrow ZnCl_{2(aq)} + H_{2(g)}$ can be increased by   A. using zinc powder instead of zinc granules   B. using dilute instead of concentrated hydrochloric acid   C. cooling the container in which the mixture is placed   D. using an alloy of zinc instead of pure zinc   E. using hydrogen chloride gas instead of hydrochloric acid solution

48) Octane number is highest in petrol containing a high proportion of   A. heptane B. octane C. 2-methylpentane D. 2, 4-dimethylpentane E. 2,2,4-trimethylpentane

49) Given the following half-cell reaction: $2Cl_{(aq)} \longrightarrow Cl_{2(g)} + 2e$, how many moles of electrons will be required to produce $1.12dm^3$ of chlorine gas at s.t.p.? (Molar volume of a gas at s.t.p. $= 22.dm^3$)   A. 100 B. 0.40 C. 0.20 D. 0.10 E. 0.01

50) Which of the following is used extensively in manufacturing photographic chemicals?   A. Aluminium chloride B. Zinc chloride C. Silver chloride D. Mercury (II) chloride E. Iron (II) chloride

51) What is concentration, in $mol\ dm^{-3}$, of a solution containing 0.10g of sodium hydroxide per $50cm^3$ of solution? (NaOH= 40) A. 0.05 B. 01 C. 0.5 D. 2 E. 5

52) The following are s-block elements except   A. lithium B. magnesium C. potassium D. calcium E. aluminium

53) Which of the following statements is not true of tin? It   A. is a p-block element B. is extracted from cassiterite, SnO_2 C. is used for protecting iron containers from corrosion D. combines with copper to form the alloy bronze E. shows two oxidation states, +2 and +4, in most of its compounds.

54) What quantity of copper will be deposited by the same quantity of electricity that deposited 9.0g of aluminium? (A = 27, Cu = 64)   A. 64g B. 32g C. 7.1g D. 6.4g E. 3.2g

55) Given the information in the table below:

Nucleus	Half-life
R	Very high
V	Not very high
S	High
T	Very low
U	Low

Which of the nuclei is the most stable?   A. U B. T C. S D. V E. R

56) The electronic configuration of two atoms X and Y are as follows: $X - 1s^2 2s^2 2p^6 3s^1$, Y- $1s^2 2s^2 2p^6 3s^2 3p^6 4s^2$

Which of the statements below is true of the position of X and Y in the Periodic Table?
    A. X belongs to group 1, Y belongs to period 2,   B. X belongs to group 1, Y belongs to period 4   C. X belongs to group 2, Y belongs to period 1
D. X belongs to group 3, Y belongs to period 2   E. X belongs to group 3, Y belongs to period 4

57) What is the concentration in $mol\ dm^{-3}$ of a 10.1g mass of potassium trioxonitate (V) in $50cm^3$ of solution? ($KNO_3 = 101$)   A. $1.0\ mol\ dm^{-3}$ B. $1.5\ mol\ dm^{-3}$ C. $3.0\ mol\ dm^{-3}$ D. $2.0\ mol\ dm^{-3}$ E. $5.0\ mol\ dm^{-3}$

58) The oxidation number of cholorine is +1 in   A. $KClO_3$ B. Cl_2O_7 C. $ZnCl_2$ D. HCl   E. NaClO

59) Catalytic hydrogenation of oils results in the production of   A. soaps   B. detergents   C. alkanes   D. butter   E. Margarine

60) A sample of orange juice is suspected to have been contaminated with a yellow dye. Which of the following methods can be used to detect the dye?   A. Decantation   B. Chromatography   C. Distillation   D. Filtration   E. Evaporation

61) The electronic configuration $1s^2 2s^2 2p^6 3s^2 3p^6$ is that of a   A. Noble gas   B. Group II element   C. Group III element   D. Group V element   E. Group VI element

62) The heat accompanying the reaction represented by the equation:
$H_2O_{(l)} ----> H_2O_{(g)}$ is described as the heat of   A. solution   B. neutralization   C. vaporization   D. sublimation   E. activation

63) Aluminium is above iron in the electrochemical series, yet iron corrodes easily on exposure to air while aluminium does not. This is because aluminium   A. has a lower density than iron   B. is a better conductor than iron   C. does not corrode spontaneously   D. forms a thin layer of inert oxide in moist air   E. forms amphoteric oxides while iron does not

64) When a solid melts and consequently boils, there is   A. a gradual increase in the average kinetic energy of the particle   B. a sudden decrease in the kinetic energy of the particles   C. no change in the average kinetic energy of the particles   D. a rapid change in the nature of bonding of the components   E. a change in the size of the fundamental particles.

65) Which of the following compounds will undergo addition reaction?   A. Ethyne   B. Butane   C. Pentane   D. Tetrachloromethane   E. Ethanol

66) The pH values of the solutions resulting from the dissolution of the oxides of elements M, N, O, P and Q in water are as indicated in the table below:

Element	pH of solution
M	3
N	5
O	6
P	7
Q	9

Which of the above elements is likely to be a metal?   A. M   B. N   C. O   D. P   E. Q

67) A catalyst which increases the rate of a chemical reaction does so by   A.   increasing the reaction pathway   B. increasing the surface area of the reactants   C. increasing the pressure on the system   D. decreasing the temperature at which the reaction occurs   E. decreasing the activation energy of the reaction

68) The gas that is liberated when iron is heated with concentrated tetraoxosulphate (VI) acid is   A. H_2S B. SO_2 C. HCl   D. SO_3 E. O_2

69) The compound of copper which is used in electroplating, dyeing, printing, wood preservation and as a fungicide is   A. copper (II) hydroxide   B. copper (II) trioxonitrate (V) pentahydrate   C. copper (II) tetraoxosulphate (VI) pentahydrate   D. copper (II) oxide   E. copper (II) trioxocarbonate (IV)

70) What is the molecular formula of a compound whose empirical formula is CH_2O and molar mass is 180?   (H = 1, C = 12, 0 = 16)   A. $C_4H_8O_2$ B. $C_4H_8O_3$ C. $C_6H_{10}O_5$ D. $C_6H_{12}O_6$ E. $C_{12}H_{22}O_{11}$

71) MO and M_2O represent the formulae of the compounds of an element. What law of chemical combination is represented?   A. Definite proportion   B. Conservation of matter   C. Multiple proportion   D. Reciprocal proportion   E. Constant composition

72) Which of the following pollutants is biodegradable?   A. Sewage   B. Plastics   C. Metal scraps   D. Lead compounds   E. Hydrogen sulphide

73) In the reaction represented by the equation: $2FeCl_3 + SO_2 + 2H_2O \rightarrow 2FeCl_2 + H_2SO_4 + 2HCl$, the oxidation number of sulphur changes from   A. +2 to +6   B. +4 to +6   C. 0 to +6   D. -2 to +4   E. +2 to +4

74) Which of the following statements explains why textraoxosulphate (VI) acid is regarded as a strong acid   A. Tetraoxosulphate (VI) acid is dibasic   B. The acid is concentrated   C. The acid is completely ionized in aqueous solution   D. Tetraoxosulphate (VI) ions are very reactive   E. The acid is highly corrosive

75) Which of the following is not correct of glucose and fructose? They   A.   are structural isomers B. reduce Fehling's solution C. can be obtained by the hydrolysis of starch D. are readily fermented by enzymes E. are soluble in water.

76) To what temperature must a gas be raised from 273K in order to double both its volume and pressure?   A. 298K B. 300K C. 546K D. 819K E. 1092K

77) Which of the following gasses are produced when dilute tetraoxosulphate (VI) acid reacts with a mixture of iron filings and iron (II) sulphide?   A.   Hydrogen and sulphur (IV) oxide   B. Hydrogen and hydrogen sulphide C. Hydrogen sulphide and sulphur (IV) oxide D. sulphur (IV) oxide and sulphur (VI) oxide E. Hydrogen sulphide and sulphur (VI) oxide.

78) If 3 moles of electrons are required to deposit 1 mole of metal M during the electrolysis of its molten chloride, the empirical formula of the metallic chloride is A. M_3Cl B. M_3Cl_2 C. MCl D. M_2Cl_3 E. MCl_3

79) Which of the following explains why trioxonitate (V) acid is not used for preparing hydrogen from metals? A. It is volatile   B. It is strongly oxidizing   C. It forms a layer of oxide on the metal D. Trioxonitrate (V) acid is soluble in water E. It renders the metal passive

80) Which of the following is normally used for drying ammonia gas?   A.   Concentrated tetraoxosulphate (VI) acid B. Calcium oxide   C. Anhydrous calcium chloride   D. Phosphorus (V) oxide E. Anhydrous copper (II) tetroxosulphate (VI).

81) Nuclear reactions can be used in the following except   A. gauging the thickness of objects B. making atomic bombs   C. Curing cancer   D. Generating electricity E. Purifying water

82) Which of the following compounds crystallizes without water of crystallization?   A. Na_2CO_3 B. $CuSO_4$ C. $MgSO_4$ D. $NaCl$ E. $FeSO_4$

83) The following metals are extracted by the electrolytic method except   A.   potassium B. calcium C. sodium D. tin   E. magnesium

84) Petrol can be obtained from diesel by   A.   distillation   B.   cracking   C.   catalysis   D. polymerization E. dehydrogenation

85) The metal extracted from cassiterite is   A. calcium B. copper C. tin D. sodium E. lead

86) In which of the following is the oxidation number of nitrogen zero? A. NH_3 B. $NaNO_3$ C. HNO_2 D. N_2 E. NOI_3

87) The separation of a mixture of calcium trioxocarbonate (IV) and sodium trioxocarbonate (IV) is most easily carried out by using the differences in their   A. physical states   B. melting points C. Rf. Values D. solubility E. boiling points

88) When air is passed through a heated tube containing finely divided copper, the component that is absorbed in the process is   A. carbon (IV) oxide   B. nitrogen C. oxygen D. water vapour E. noble gases

89) What is the amount (in mole) of sodium trioxocarbonate (IV) in 5.3g of the compound? (Na_2CO_3 = 106) A. 0.05 B. 0.10 C. 0.20 D. 0.50 E. 2.00

90) Which of the following elements can form more than one acidic oxide?   A. Hydrogen B. Sulphur C. Carbon D. Aluminium E. Iron

91) $CH_2 = CH_2 + H_2SO_4 -----> CH_3-CH_2-OSO_3H$. The reaction illustrated by the equation is an example of   A. substitution B. oxidation C. reduction D. addition E. displacement

92) Brass is an alloy of copper and   A. iron B. chromium C. Zinc D. tin   E. magnesium

93) The products of the electyrolysis of dilute sodium chloride solution with pantinum electrodes are   A. hydrogen and oxygen B. oxygen and chlorine C. chlorine and water D. sodium amalgam and chlorine E. sodium hydroxide and water

94) Which of the following statements is not correct? Cathode rays A. emerge at right angles to the cathode B. travel in straight lines C. are deflected away from negative plates D. are very light E. are positively charged

95) In which of the following are the substances arranged in their correct order of increasing melting point?   A. sodium chloride, sulphur, diamond   B. Diamond, sodium chloride, sulphur C. sulphur, sodium chloride, diamond D. Diamond, sulphur, sodium chloride E. Sulphur, diamond, sodium chloride.

96) A compound which liberates carbon (IV) oxide from a hydrogentrioxocarbonate (IV) could have the molecular formula A. C_2H_5OH B. C_3H_4 C. HCOOH D. H_2O_2 D. C_6H_6

97) $CH_{4(g)} + 2O_{2(g)} -----> 2H_2O_{(l)} + CO_{2(g)}$ $\triangle H$ = -890KJ mol^{-1}. $\triangle H$ in the reaction represented by the equation above is the enthalpy of A. formation   B. combustion C. solution D. activation E. neutralization

98) Which of the following modes of motion is exhibited by the particles of a solid?   A. vibrational and random motion B. Vibrational and translational motion C. Translational and random motion D. Vibrational motion only E. Translational motion only

99) Consider the reaction represented by the equation below: $H_{2(g)} + I_{2(s)} \longrightarrow 2HI_{(g)}$ $\triangle H$ is negative. Which of the following takes place when the temperature of the reaction vessel is decreased? A. The gases condense B. The yield of hydrogen increases C. The concentration of the reactants remain constant D. More of the hydrogen iodide decomposes E. The yield of hydrogen iodide increases

100) Esterification of propane 1, 2, 3 – triol and unsaturated higher carboxylic acids will produce. A. fats B. soap C. methyl-propanoate D. alkanols E. Oils

101) The ratio of the number of protons to the number of electrons in the atom $_{15}X^{3-}$ is A. 1:2 B. 5:6 C. 2:1 D. 1:5 E. 5:1

102) When a mixture of caustic soda and olive oil is heated, it produces A. fats B. soap C. margarine D. alkanols E. Butter

103) The ratio of reactants to products is 1 : 3 : 2 in the reaction represented by the equation: $N_{2(g)} + 3H_{2(g)} \longrightarrow 2NH_{3(g)}$. Which of the following law is demonstrated by this? A. Boyle's law B. Law of multiple proportion   C. Gay Lussac's law D. Law of constant composition   E. Avogadro's law

104) In the redox reaction represented by the following equation:
$Cu^{2+}_{(aq)} + Zn_{(s)} \longrightarrow Cu_{(s)} + Zn^{2+}$ A. the oxidation number of copper increases B. Copper (II) ions are reduced to copper atoms C. zinc atoms are reduced to zinc ions D. copper (II) ions donate electrons to zinc atoms

105) When starch undergoes complete enzyme-catalyzed hydrolysis, the resulting product is A. glucose B. maltose C. sucrose D. fructose E. cellulose

106) Which of the following is not a direct petroleum product? A. Methane B. Ethanol C. Petrol D. Vaseline E. Kerosene

107) How many principal electronic shells has an atom whose electronic configuration is shown below? $1s^2 2s^2 2p^6 3s^1$ A. 1 B. 2 C. 3 D. 4 E. 11

108) Which of the following properties is not characteristic of electrovalent compounds?   A. Formation of crystalline solids B. Solubility in polar solvents C. High vapour pressures D. High melting and boiling points E. Ability to conduct electricity in aqueous solution

109) Compounds that have the same molecular formular but different structures are said to be A. allotropic B. polymorphic C. polymeric D. isomeric E. isotopic

110) In which of the following crystals are the particles held by van der Waal's forces only? A. Sodium chloride B. Ice C. Diamond D. Iodine E. Quartz

111) A current of 4.0 amperes was passed through copper (II) tetraoxosulphate (VII) solution for one hour using copper electrodes. What was the mass of copper deposited? (Cu = 64, 1F = 96500 C) A. 6.4 B. 9.6 C. 3.2 D. 4. 8 E. 12.8

112) When a crystal was added to its solution, it did not dissolve and the solution remained unchanged, showing that the solution was A. concentrated B. supersaturated C. unsaturated D. colloidal E. saturated

113) An organic compound which reacts readily with bromine to form a compound with the formula $CH_3CHBrCH_2Br$ is A. ethane B. propane C. butane D. propene E. propyne

114) Water from a river contaminated by alkali waste will have a pH of about
A. 1 B. 3 C. 5 D. 7 E. 9

115) Hydrogen is used for the following except A. manufacturing of ammonia B. synthesis of hydrochloric acid C. extinguishing fire D. conversion of coal to petrol E. production of margarine

116) Which of the following is suitable for determining different isotopes present in an element which exhibits isotopy? A. Sensitive weighing balance B. Cathode ray tube C. Mass spectrometer D. Geiger muller counter E. Eudiometer

117) When steam is passed over white-hot coke, the products are A. carbon(IV) oxide and nitrogen B. carbon (IV) oxide and hydrogen C. carbon (II) oxide and nitrogen D. carbon (II) oxide and hydrogen E. carbon (IV) oxide and steam

118) The maximum number of electrons that can be accommodated in the shell having the principal quantum number 3 is A. 3 B. 9 C. 10 D. 18 E. 32

119) Methanol is obtained from wood by A. esterification B. bacteria decomposition C. combustion D. fractional distillation E. destructive distillation

120) Study carefully the reaction represented by the equation below
$2H_2O_{2(l)} \longrightarrow O_{2(g)} + 2H_2O_{(l)}$. Which of the following will not increase the reaction rate? A. Heating the hydrogen peroxide B. Adding a pinch of MnO_2 to the reactant C. Increasing the concentration of the H_2O_2 D. Adding water to the reactant E. Exposing the reactant to sunlight.

121) Which of the following processes is a physical reaction? A. Electrolysis B. Hydrolysis C. Allotropic change D. Neutralization E. Corrosion

122) The following acids are monobasic except   A. methanoic acid   B. dioxonitrate III acid   C. ethanedioic acid D. oxochlorate (I) acid E. hydrobromic acid

123) The reactions of ethyne are mainly   A. substitution reactions   B. addition reactions   C. addition polymerization   D. catalytic hyrodgenation E. catalytic halogenations

124) When air is passed through potash and then pyrogallol, the components remaining are noble gases, A. oxygen and carbon (IV) oxide   B. oxygen and nitrogen   C. nitrogen and water vapour D. water vapour and carbon (IV) oxide E. oxygen and water vapour

125) The halide used widely in photography is   A. ammonium chloride B. calcium chloride C. silver bromide D. sodium bromide E. aluminium fluoride

126) The scum formed when soap is mixed with hard water could be   A. calcium trioxocarbonate (IV) B. propane-1,2,3-triol   C. calcium hydrogentrioxocarbonate (IV)   D. magnesium tetraoxosulpahte (VI) E. calcium stearate

127) What is the quantity of electricity produced when a current of 0.5A is passed for 5 hours 45 mins? (1F = 96500C)   A. 0.11F   B. 0.12F   C. 0.22F   D. 1.1F   E. 2.2F

128) "The rate of a reaction is proportional to the number of effective collisions occurring per second between the reactants". This statement is associated with   A. kinetic theory   B. rate law   C. atomic theory D. collision theory E. gas laws

129) In the reaction represented by the following equation, $2H_2S_{(g)} + SO_{2g}$ -----> $2H_2O_{(l)} + 3S_{(S)}$, SO_2 is acting as   A. a reducing agent   B. an oxidizing agent   C. a dehydrating agent   D. a bleaching agent   E. a precipitating agent

130) What is the change in the oxidation number of phosphorus in the reaction represented by the following equation? $4P_{(s)} + 5O_{2(g)}$ -----> $2P_2O_{5(g)}$
A. 0 to +2 B. 0 to +5   C. +4 to +2   D. +4 to +5   E. +4 to +7

131) Which of the following compounds will not decompose when heated strongly?   A. $NaHCO_3$ B. K_2CO_3 C. $MgCO_3$ D. $Ca(HCO_3)_2$ E. $ZnCO_3$

132) The following salts are readily soluble in water except   A. Na_2CO_3 B. $Pb(NO_3)_2$ C. KCl   D. $CuCO_3$ E. $FeSO_4$

133) When sucrose is warmed with Fehling's solution   A. a silver mirror is produced   B. the solution turns milky   C. a brick-red precipitate is formed   D. there is no precipitate E. a blue-black coloration is produced

15

134) Which of the following gases will have the highest rate of diffusion under the same conditions? ($H = 1, C = 12, O = 16, S = 32, Cl = 35.5$)
A. O_2 B. Cl_2 C. HCl D. H_2S E. CO_2

135) The ionic radii of metals are usually A. greater than their atomic radii B. unaffected by the charge on the ion. C. less than their atomic radii. D. greater than those of non-metals E. the same as their atomic radii

136) The formula $(CH_3)_3$ COH is that of a A. polyhdric alkanol B. Secondary alkanol C. tertiary alkanol D. primary alkanol E. trihydric alkanol

137) Which of the following compounds is not a raw material for the manufacture of plastics? A. Ethyne B. Ethane C. Monochloroethene D. Propene E. Butadiene

138) The energy required to remove the most loosely bound electron from an atom in the gaseous state is known as the A. bond energy B. ionization energy C. potential energy D. activation energy E. kinetic energy

139) Vulcanization of rubber A. stops the growth of fungus on the rubber B. increases the solubility of the rubber C. decrease the elasticity of the rubber D. hardens the rubber through cross-linkage E. removes sulphur from the rubber

140) If a reaction is said to be exothermic, which of the following statements is a correct deduction from the information? A. The reaction vessel gets hotter as the reaction proceeds B. $\triangle H$ for the reaction is positive C. The rate of the reaction increases with time D. The activation energy of the reaction is high   E. The reaction vessel gets cooler as the reaction proceeds

141) Which of the following statements is not correct about esterification? A. It is a slow reaction B. the process is reversible C. it is similar to hydrolysis D. it is catalysed by acids E. the products usually have a fruity smell

142) Which of the following statements is not correct? The 4s orbital   A. is defined by the quantum number $l = 0$ B. is of higher energy than the 3d orbital C. contains a maximum of two electrons D. is spherical about the nucleus E. is filled before the 4p orbital

143) Which of the following pH values is likely to be that of a slightly alkaline solution? A. 2 B. 5 C. 7 D. 8 E. 13

144) Which of the following minerals contains fluorine as one of its constituent elements? A. Cryolite B. Bauxite C. Potash alum D. Kaolin E. Mica

145) The product of the reaction between propanoic acid and ethanol is   A. propylmethanoate   B. ethylethanoate   C. methylpropanoate   D. propylethanoate   E. ethylpropanoate

146) Which of the following groups contains entirely linear molecules?
A. H_2, NH_3, O_2 B. CO_2, NH_3, N_2 C. H_2, CH_4, N_2 D. CO_2, H_2, N_2,   E. CH_4, O_2, CO_2

147) What is the value of n in the following equation? $XO_4^- + 8H^+ + ne^- \longrightarrow X^{2+} + 4H_2O$ A. 2   B. 3   C. 4   D. 5   E. 6

148) A dye is suspected to have contaminated a lollipop. Which of the following is the best method by which the contaminant may be isolated?   A. Fractional distillation   B. Recrystallization   C. Filtration   D. Paper chromatography   E. Evaporation

149) Which of the following accounts for the difference in the mode of conduction of electricity by metals and aqueous salt solutions?   A. Electrons are present in metals but not in salt solutions   B. Metals are conductors while salts are electrolytes   C. Electricity is carried by mobile electrons in metals but by ions in aqueous salt solution   D. salts ionize in aqueous solutions while metals do not   E. Metals are reducing agents while salts are not

150) Which of the following separation is routinely applied in the petroleum industry   A. Filtration   B. Chromatography   C. Evaporation   D. Fractional crystallization   E. Fractional distillation

151) Starch undergoes complete hydrolysis to produce   A. maltose   B. lactose   C. fructose   D. glucose   E. sucrose

152) Which of the following solids has a hexagonal structure?   A. Diamond   B. Iodine   C. Sulphur   D. Graphite   E. iron

153) Trioxosulphate (IV) acid is not stored for a long period because it   A. Is a weak dibasic acid   B. has bleaching properties   C. is unstable and easily decomposed   D. smells strongly of sulphur (IV) oxide   E. can act as a germicide

154) Which of the following statements is not correct about the compound represented below?
COOH   A. Its basicity is 2.   B. It is a di-carboxylic acid
| C. Its boiling point is lower than that of the corresponding alkane
COOH   D. It reacts with alkanols under suitable conditions
                E. It produces effervescence with saturated $NaHCO_3$ solution

155) Alkanes are used mainly   A. In the production of plastics   B. as domestic and industrial fuels   C. in the textile industry   D. in the hydrogenation of oils   E. as fine chemicals

156) Which of the following solids will leave a black residue after being heated strongly? A. Iron (II) tetraoxosulphate (VI) B. Lead (II) trioxocarbonate (IV) C. Copper (II) trioxonitrate (V) D. Calcium trioxocarbonate (IV) E. Magnesium trioxocarbonate (IV)

157) The properties of electrovalent compounds include the following except
A. high melting point and boiling point B. conduction of electricity in the molten state C. high volatility at room temperature D. ionization in aqueous solution E. decomposition of their solutions by electric current

158) Which of the following pairs illustrates isotopy? A. But-1-ene and but-2-ene B. Ortho-hydrogen and parahydrogen C. Oxygen and ozone D. Hydrogen and deuterium E. Alpha-particle and beta-particle

159) Carbon is often deposited in the exhaust-pipe of cars because of the A. presence of carbon in petrol B. dehydration of petrol C. incomplete combustion of petrol D. presence of additives in petrol E. contamination of petrol with diesel

160) What does X stand for in the following expression? $X = -Log_{10}(H^+)$ A. Enthalpy change ($\triangle H$) B. Equilibrium constant (K) C. Solubility product D. Degree of acidity (pH) E. Ionic product of water

161) Sulphur burns in air to form A. an acidic oxide B. a basis oxide C. an amphoteric oxide D. a neutral oxide E. a mixed oxide

162) Chlorine is used in water treatment as A. a germicide B. a decolorizing agent C. an antioxidant D. a coagulating agent E. an aerating agent

163) Which of the following pieces of apparatus can be used for drawing a current of air through a liquid? A. Wash bottle B. Aspirator C. Capillary tube D. Thistle funnel E. Pipette

164) What amount of copper will be deposited if a current of 10A was passed through a solution of copper (II) salt for 965 seconds? (1F = 96500 C) A. 0.005 mole B. 0.025 mole C. 0.05 mole D. 1.00 mole E. 1.05 mole

165) Some precious stones such as ruby and sapphire consist of aluminum oxide coloured by traces of the oxides of A. group IV elements B. group I metals C. transition metals D. alkaline earths metals E. the halogens

166) The following gases decolorize bromine water except A. C_2H_6 B. C_2H_4 C. C_2H_2 D. C_3H_4 E. C_3H_6

167) What volume of distilled water should be added to $400cm^3$ of 2.0 mole dm^{-3} H_2SO_4 to obtain 0.20 mole dm^{-3} of solution? A. $600cm^3$ B. $800cm^3$ C. $1,000cm^3$ D. $3,600cm^3$ E. $4,000cm^3$

168) Which of the following indicates the correct increasing order of oxidizing power of the halogens? A. F < Cl < Br < I B. I < Br < Cl < F C. I < Cl < Br < F D. Br < I Cl < F E. Cl < F < I < Br

169) Which of the following substances is a peroxide? A. Na_2O_2 B. CuO C. Pb_3O_4 D. Fe_2O_3 E. Al_2O_3

170) The component of air that is removed when air is bubbled into alkaline pyrogallol solution is A. carbon (IV) oxide B. oxygen C. water vapour D. nitrogen E. helium

171) Which of the following will not affect the degree of hardness in a sample of temporarily hard water? A. Addition of washing soda B. Adding of soap C. Treatment with ion exchange resin D. Filtration E. Distillation

172) Examples of polymers include the following except A. starch B. nylon C. wool D. Perspex E. glass

173) When the trioxonitrate (V) salt of an alkali metal Y is heated, the formula of the residue is A. Y_2O B. YNO_2 C. Y_2O_3 D. $Y(NO_2)_2$ E. YO_2

174) Producer gas is a mixture of A. oxygen and hydrogen B. hydrogen and nitrogen C. nitrogen and oxygen D. nitrogen and carbon (II) oxide E. hydrogen and carbon (II) oxide

175) An alkene may be converted to an alkane by A. halogenations B. hydrolysis C. dehydration D. hydrogenation E. Decomposition

176) Which of the following compounds has the highest ionic character? A. $KCl_{(s)}$ B. $BeCl_{2(s)}$ C. $HCl_{(g)}$ D. $AlCl_{3(s)}$ E. $CaCl_{2(s)}$

177) The following gases are diatomic except A. nitrogen B. helium C. hydrogen D. fluorine E. oxygen

178) The product of the reaction between ethanol and excess acidified $K_2Cr_2O_7$ is A. ethanal B. ethylethanoate C. ethanoic acid D. ethyne E. ethanedioic acid

179) What does the following equation illustrate?
$$^{238}_{92}U \longrightarrow\ ^{234}_{90}Th\ +\ ^{4}_{2}He$$
A. Nuclear fission B. Nuclear fusion C. Artificial radioactivity D. Spontaneous disintegration E. Thermal decomposition

180) When ammonium chloride is dissolved in water in a test tube, the tube feels cooler showing that A. the solution is unsaturated B. sublimation has occurred C. ammonia gas is evolved D. the process is endothermic E. condensation has occurred

181) Zinc displaces copper from an aqueous solution of copper (II) salt because   A. copper is a transition element   B. copper is a moderately reactive metal   C. zinc and copper have reducing properties   D. zinc is more reactive than copper   E. zinc reacts with both acids and alkalis

182) The alloy used for metal work and plumbing contains   A. lead and tin   B. Iron and carbon   C. copper and tin   D. aluminium and copper   E. lead and antimony

183) The components of universal indicator solution can best be separated by
A. Chromatography   B. Filtration   C. Evaporation   D. crystallization   E. fractional distillation.

184) The number of replaceable hydrogen atoms in one molecule of an acid indicates its   A. basicity   B. acidity   C. alkalinity   D. reactivity   E. pH value

185) Catalytic hydrogenation of alkenes produces compounds with the general formula   A. $C_nH_{2(n+1)}OH$ B. C_nH_{2n+1} C. C_nH_{2n-2} D. C_nH_{2n+2} E. $C_x(H_2O)_y$

186) A measure of the degree of disorder in a chemical system is known as the   A. enthalpy   B. free energy   C. activation energy   D. entropy   E. equilibrium

187) In any chemical reaction, the total mass of the products is always equal to that of the reactants. This is statement of the law of   A. definite proportion   B. conservation of matter   C. multiple proportions   D. reciprocal proportions   E. constant composition

188) Which of the following is not a property of magnesium oxide?   A. High melting point   B. Dissolution in polar solvents   C. Presence of ionic bonds   D. Possession of crystal lattice   E. low binding energy

189) $^{235}_{92}U + ^{1}_{0}n \longrightarrow ^{145}_{56}Ba + ^{88}_{36}Kr + 3^{1}_{0}n + energy$
The reaction represented by the equation above is   A. an alpha particle emission   B. an endothermic reaction   C. a thermal ionization   D. nuclear fission   E. a nuclear fusion

190) Which of the following would increase the rate of reaction of chips of a metal M as shown in the equation below? $M_{(s)} + 2HCl_{(aq)} \longrightarrow MCl_{2(aq)} + H_{2(g)}$
A. Diluting the hydrochloric acid   B. Adding some distilled water   C. Decreasing the temperature   D. Grinding the metal into powder   E. increasing the pressure

191) The oxidation number of phosphorus in PO_4^{3-} is   A. +1   B. +2   C. +3   D. +4   E. +5

192) The number of atoms in 1.2g of carbon – 12 isotope is   A. 1.00×10^{22} B. 6.02×10^{22} C. 1.20×10^{23} D. 1.00×10^{24} E. 6.02×10^{24}

193) Tetraoxosulphate (VI) acid is described as strong acid because it is highly   A. corrosive   B. concentrated   C. reactive   D. soluble in water   E. ionized in water

194) The following will decolourize acidified potassium tetraoxomanganate(VII) solution except   A. sulphur(IV) oxide   B. but-3-ene   C. pentane   D. hydrogen sulphide   E. propyne

195) Polyatomicity is illustrated by molecules of   A. sulphur   B. carbon (IV) oxide   C. noble gases   D. liquefied oxygen   E. liquefied nitrogen

196) Calcium is usually extracted by the electrolysis of its   A. trioxonitrate (V)   B. tetraoxosulphate (VI)   C. oxide   D. hydride   E. chloride

197) When naphthalene on heating changes from the solid state directly to the gaseous state it undergoes   A. evaporation   B. sublimation   C. decomposition   D. combustion   E. ionisation

198) Which of the following occurs when an aqueous solution of sodium hydroxide is electrolysed using graphite electrodes?   A. sodium metal is produced at the anode   B. sodium amalgam is formed at the cathode   C. oxygen gas is produced at the anode   D. the graphite anode dissolves   E. the resulting solution becomes acidic

199) Glucose can be obtained from starch by   A. hydrogenation   B. dissociation   C. hydrolysis   D. dialysis   E. dehydration.

200) How many faradays of electricity are required to liberate 9g of aluminium? (Al = 27)   A. 0.1   B. 0.3   C. 1.0   D. 2.7   E. 3.0

201) When water is dropped on calcium carbide, the gaseous product is an   A. alkane   B. alkene   C. alkyne   D. alkanol   E. alkanal

202) $Mg_{(s)} + 2HCl_{(aq)} \longrightarrow MgCl_{2(aq)} + H_{2(g)}$. From the equation above, what mass of hydrogen would be produced if 12.0g of magnesium reacted completely with dilute hydrochloric acid? (H = 1, Mg = 24).   A. 1.0g   B. 2.0g   C. 6.0g   D. 12.0g   E. 24.0g.

203) The position of an element in the Periodic Table is determined by   A. its density   B. its atomic radius   C. its relative atomic mass   D. the number of protons in its atom   E. the number of neutrons in its atom

204) In a chemical reaction, the reacting species possess energy of motion known as   A. potential energy   B. free energy   C. bond energy   D. kinetic energy   E. hydration energy

205) The following compounds will decompose on heating except   A. $NaHCO_3$ B. K_2CO_3 C. $MgCO_3$ D. $Ca(HCO_3)_2$ E. $ZnCO_3$

206) What is the molar mass of an alkyne with the formula C_xH_{14}? (H = 1, C = 12)
A. 86g B. 92g C. 98g D. 110g E. 112g

207) What is the role of hydrogen sulphide gas in the reaction represented by the following equation? $H_2SO_{4(aq)} + 3H_2S_{(g)} -----> 4S_{(s)} + 4H_2O_{(l)}$ A. Reducing agent B. A bleaching agent C. A dehydrating agent D. A sulphonating agent E. An oxidizing agent

208) Alkanols have unexpectedly high boiling points relative to their molar masses because of intermolecular   A. hydrogen bonding B. metallic bonding C. covalent bonding D. ionic bonding E. van der Waals' forces

209) Sodium chloride is used in the following processes except in the   A.   treatment of polluted water B. preservation of food C. separation of soap from glycerol D. manufacture of chlorine E. manufacture of caustic soda

210) The gaseous product formed when ammonia is passed over heated copper (II) oxide is   A. oxgen B. nitrogen C. hydrogen D. nitrogen (I) oxide E. nitrogen (II) oxide

211) Which of the following is not a halogen? A. Silicon B. Fluorine C. Astatine D. Bromine E. Iodine

212) Which of the following is likely to have the highest degree of hardness? A. Carbonated water B. Distilled water C. Acidified water D. Rain water E. Lime water

213) Destructive distillation of coal means   A. heating coal in plentiful supply of air B. burning coal in air to produce water and carbon (IV) oxide C. heating coal in the absence of air D. heating coal in limited supply of air E. separating coal into various allotropes of carbon

214) When magnesium ribbon burns in air, the products are   A. magnesium nitride and magnesium oxide   B. magnesium nitride and soot   C. magnesium oxide and steam D. magnesium oxide and carbon (II) oxide E. magnesium nitride and carbon (II) oxide

215) The following compounds contain the same type of bonds except   A. sodium chloride B. hydrogen chloride C. magnesium chloride D. potassium chloride E. lithium chloride

216) Given the element $^{40}_{20}Y$, it can be deduced that Y has   A.   60 neutrons   B. an atomic number of 20 C. a mass number of 60 D. 40 electrons E. 40 protons

217) All pure samples of the same chemical compound contain elements combined in the same proportion by mass is a statement of   A. the law of conservation of matter B. the law of constant composition   C. the law of multiple proportion   D. Gay Lussac's law   E. Avogadro's law

218) If an element **X** with electronic configuration 2, 8, 3, combines with another element **Z** with electronic configuration 2, 8, 6, the compound formed will have formula   A. XZ   B. XZ_2 C. X_2Z D. X_2Z_3 E. X_3Z_2

219) A beta particle is represented as   A. ^1_0n B. $^0_{-1}e$ C. $^0_{+1}E$ D. ^1_1H E. ^4_2He

220) If a solid has a low melting point and dissolves readily in benzene, it would probably   A. contain strong electrostatic forces of attraction   B. conduct electricity in the molten state   C. dissolve in water   D. have covalent bonding   E. absorb moisture on exposure to air

221) Which of the following molecules is linear in shape?   A. CH_4 B. H_2O C. NH_3 D. H_2S E. Cl_2

222) Which of the following cannot be deduced from the electronic configuration of a transition metal? A. Possession of magnetic property   B. Ability to form complex ions   C. Position in the Periodic Table   D. Variable oxidation states   E. Physical properties of the metal

223) What is the percentage by mass of copper in copper(I) oxide (Cu_2O)? [O=16; Cu=64]   A. 88.9%   B. 80.0%   C. 66.7%   D. 20.0%   E. 11.1%

224) A weak acid is one which   A. is not corrosive   B. is slightly ionized in water   C. does not produce salts with alkalis   D. does not conduct an electric current in aqueous solution   E. forms acid salts with bases.

225) Which of the following species is always present in acidified water?   A. NH^+_4 B. O^{2-} C. HCl   D. H_3O^+ E. HNO_3

226) A positive brown ring test indicates the presence of   A. NO_3 B. Fe^{3+} C. SO^{2-} D. Cu^+ E. NO^-_2

227) Consider the reaction represented by the equation below: XO + YO -----> X + YO_2. In the reaction, YO acts as   A. an acidic oxide   B. a basic oxide   C. a reducing agent   D. a weak base   E. an oxidizing agent

228) Sodium chloride cannot conduct electricity in the solid state because it   A. is a normal salt   B. is highly soluble in water   C. is an electrovalent compound   D. does not have any effect on litmus   E. does not contain mobile ions

229) What mass of copper will be deposited by the liberation of Cu^{2+} when 0.1F of electricity flows through an aqueous solution of a copper (II) salt?   [Cu = 64]   A. 64g   B. 32g   C. 12.8G   d. 6.4G   E. 3.2g

23

230) Alums are classified as   A. simple salts   B. acid salts   C. anhydrous salts D. double salts   E. basic salts

231) Aqueous sodium trioxocarbonate (IV) solution in alkaline because the salt is
A.   Hydrolysed in water   B. formed from a strong base   C. fully ionized in water
D. not decomposed by heat   E. a strong electrolyte

232) $H_3O^+_{(aq)} + OH^-_{(aq)} -----> 2H_2O_{(l)}$. The heat change accompanying the process represented by the equation above is the heat of   A. neutralization B. formation   C. solution D. dilution E. hydration

233) The following statements about graphite are correct except that it
A. is used as a lubricant   B. has a network structure   C. contains mobile free electrons   D. is a good conductor E. is an allotrope of carbon

234) Which of the following hydrocarbons is unsaturated?   A. Ethane B. Benzene C. Propane
D. 2 – methylbutane   E. 2, 2, 4 – trimethylpentane

235) The following can be obtained directly from the destructive distillation of coal except   A. ammoniacal liquor B. coke C. producer gas D. coal gas E. coal tar.

236) Fructose and glucose are classified as carbohydrates because they   A.   are   crystalline white solids   B. are energy-giving substances   C. are naturally occurring organic compounds D. contain carbon, oxygen and hydrogen E. conform to the general formula $C_x(H_2O)_y$

237) The gas produced when a mixture of sodium propanoate and soda lime is heated is   A. methane B. pentane C. ethane D. butane E. ethyne

238) In which of the following processes are larger molecules broken down into smaller molecules?
A. Vulcanization of rubber B. Hydrogenation of palm oil   C. Hydrolysis of starch D. Polymerization E. Esterification

239) Which of the following is an alloy of mercury?   A.   Stainless   steel   B. Soft   solder   C. Coinage bronze D. Amalgam E. Duralumin

240) The function of limestone in the extraction of iron in the blast furnace is   A. removal of the earthly impurities   B. decomposition of the iron ore   C. conversion of iron (III) to iron (II) compounds   D. generation of heat for the process E. conversion of coke to carbon (II) oxide

241) Metals which react with steam only when they are red-hot include   A. copper B. sodium C. calcium D. gold E. iron

24

242) Copper (II) tetraoxosulphate (VI) is often added to swimming pools because it A. prevents the growth of algae B. coagulates suspended particles C. neutralizes dissolved gases in water D. reacts with any excess chlorine present E. increases the amount of dissolved oxygen

243) The following salts dissolve readily in cold water except A. $CaCl_2$ B. $PbSO_4$ C. $(NH_4)_2SO_4$ D. Na_2CO_3 E. Na_2SO_3

244) Which of the following compounds is not decomposed by heat? A. Sodium trioxocarbonate(IV) B. Ammonium trioxocarbonate(IV) C. Sodium hydrogen trioxocarbonate(IV) D. Calcium trioxocarboante (IV) E. Calcium hydrogentrioxocarboante(IV).

245) What is the amount (in moles) of hydrogen gas that would be produced if 0.6 mole of hydrochloric acid reacted with excess zinc according to the following equation? $Zn_{(s)}$ + $2HCl_{(aq)}$ --- $> ZnCl_{2(aq)} + H_{2(g)}$
A. 0.1 mole B. 0.2 mole C. 0.3 mole D. 1.0 mole E. 2.9 moles

246) Which of the following oxides is amphoteric? A. Na_2O B. Fe_2O_3 C. Al_2O_3 D. CaO E. CuO

247) A sample of air that has been passed through caustic potash will A. contain more oxygen than nitrogen B. rekindle a glowing splint C. have no action on lime water D. contain no water vapour E. be free of noble gases

248) The existence of two or more forms of the same element in the same physical state is known as A. allotropy B. resonance C. hybridization D. isotopy E. isomerism

249) When an atom gains an electron, it becomes A. chemically inactive B. negatively charged C. oxidized D. a cation E. a complex ion

250) The alkali metals exhibit similar chemical properties because A. they occur in the combined state B. they have the same number of valence electrons C. they form crystalline salts D. their salts are soluble in water E. they are highly reactive

251) What is the likely formula of compound formed between element M in group two and element X in group seven? A. M_7X_2 B. M_2X C. M_2X_7 D. MX_2 E. MX_6

252) "Equal volumes of all gases at the same temperature and pressure contain the same number of molecules" is an expression of A. Charles's Law B. Boyle's Law C. Graham's Law D. Avogadro's Law E. Gay Lussac's Law

253) A solution of pH 7 is A. acidic B. neutral C. concentrated D. dilute E. saturated

254) Amphoteric oxides are oxides which   A. react with water to form acids B. react with water to form alkali C. show neither acid nor basic properties D. react with both acids and alkalis E. contain high proportion of oxygen

255) The following acids are monobasic except A. trioxonitrate (V) acid B. hydrochloric acid C. ethanoic acid D. tetraoxophosphate (V) acid E. dioxonitrate (III) acid

256) An arrangement of two different metals in aqueous solutions of their slats to produce an electronic current is known as   A. electrochemical cell B. activity series C. thermocouple D. voltameter E. galvanometer

257) In which of the following compounds is the oxidation number of nitrogen equal to +3? A. NO_2 B. N_2O C. NO D. HNO_2 E. HNO_3

258) Which of the following statements is not correct about electrolysis? A. Reduction occurs at the anode B. Anions migrate to the anode C. Positive ions migrate to the cathode D. Concentration affects the discharge of ions E. Electrolytes conduct electric current

259) Which of the following is not correct about a catalyst? It A. remains unchanged chemically at the end of a reaction B. helps to establish equilibrium faster in reversible reaction C. can start a chemical reaction which will normally not take place D. is usually specific in its action E. alters the rate of chemical reaction

260) What does X represent in the following equation? $H_{2(g)} + \frac{1}{2}O_{2(g)} \longrightarrow H_2O_{(l)}$, $\triangle H = xkJ$ A. Bond energy B. Activation energy C. Ionization energy D. Enthalpy of neutralization E. Enthalpy of formation

261) If the third member of a homologous series is C_3H_8, the fifth member will be A. C_5H_8 B. C_5H_9 C. C_5H_{10} D. C_5H_{11} E. C_5H_{12}

262) Which of the following exhibits resonance? A. Benzene B. Butane C. Pentene D. Octane E. Hexene

263) On exposing palm wine to air for some days, it becomes sour owing to the conversion of A. glucose to ethanol B. glucose to gluconic acid C. ethanol to ethanoic acid D. ethanol to ethanal E. palm wine to palmitic acid

264) When acidified $KMnO_4$ solution is decolorized by ethene, the gas acts as A. a straight-chain hydrocarbon B. a saturated hydrocarbon C. a reducing agent D. a dehydrating agent E. an oxidizing agent

265) What amount of hydrogen will be required for complete hydrogenation of one mole of pent-3-yne? A. 1 mole B. 2 moles C. 3 moles D. 5 moles E. 6 moles

266) When excess carbon(IV) oxide is passed into lime water, the turbidity produced initially, disappears due to the formation of A. calcium oxide B. calcium carbide C. calcium hydrogentrioxocarbonate (IV) D. trioxocarbonate(IV) acid E. calcium trioxocarbonate (IV)

267) Which of the following is a disaccharide? A. Fructose B. Starch C. Cellulose D. Sucrose E. Glycogen

A compound X was boiled with concentrated hydrochloric acid to produce a sweet tasting substance.

Use the information above to answer question 268 and 269 below.

268) Compound X was probably A. rubber B. polyethene C. nylon D. cellulose E. ethanol

269) What type of reaction was involved between compound X and concentrated hydrochloric acid? A. Polymerization B. Fermentation C. Hydrolysis D. Neutralization E. Esterification

270) The component of air that can be removed by alkaline pyrogallol solution is A. oxygen B. nitrogen C. water vapour D. carbon (IV) oxide E. noble gases

271) During water treatment for town supply, water is passed through layers of sand beds in order to A. soften the water B. filter the water C. decolorize the water D. destroy the germs in the water E. coagulate colloidal particles in the water

272) The cleansing action of soap in hard water is not satisfactory because soap A. forms insoluble calcium and magnesium salts B. is made from fats and oils C. has an organic component D. alters the surface tension of water E. is a sodium salt of higher carboxylic acids

273) Nitrogen is prepared on a large scale by the A. fractional distillation of liquefied air B. decomposition of ammonium dioxonitrate (III) C. electrolysis of brine D. Haber process E. contact process

274) Which of the following metals will be the most suitable for use where lightness and resistance to corrosion are of importance? A. Lead B. Copper C. Iron D. Calcium E. Aluminium

275) The following are properties of transition metals except A. variable oxidation states B. tendency to form complex ions C. formation of coloured ions D. ability to act as catalyst E. low melting points

276) The following salts will produce a gas on reacting with hydrochloric acid except A. $CuSO_4$ B. $CaCO_3$ C. FeS D. Na_2SO_3 E. K_2CO_3

277) The phenomenon observed when dust particles collide randomnly in a beam of sunlight is known as A. Tyndal effect B. diffusion C. osmosis D. Brownian movement E. dialysis

278) The mass number of an atom of an element is the sum of its A. protons and electrons B. charge and electrons C. protons and neutrons D. electrons and neutrons E. neutrons and atomic number

279) What is the most probable group of an element which is a soft, silvery white solid and reacts violently with water? A. Group 0 B. Group 1 C. Group 4 d. Group 6 E. Group 7

280) In linear molecules, the bond angle is A. 90° B. 104° C. 109° D. 120° E. 180°

281) A sodium atom and a sodium ion have the same A. number of neutrons B. vapour density C. volume D. temperature E. concentration

282) If the atomic number of an element X is 11 and that of nitrogen is 7, the most likely formula of the nitride of X is A. X_3N B. XN_3 C. X_3N_2 D. N_2X E. NX_2

283) An increase in the pressure of a gas results in a decrease in its A. mass B. vapour density C. volume D. temperature E. concentration

284) An acid is a substance which in the presence of water produces A. salts B. oxygen C. effervescence D. hydroxonium ions E. hydrogen gas

285) The loss of molecules of water of crystallization to the atmosphere by some crystalline salts is known as A. efflorescence B. effervescence C. phosphorescence D. fluorescence E. deliquescence

286) Which of the following ions will migrate to the cathode during electrolysis?
A. Zinc ions B. Chloride ions C. Sulphide ions D. Tetraoxosulphate (VI) ions E. Trioxonitrate (V) ions

287) What quantity of electrons (in mole) is lost when one mole of iron (II) ions is oxidized to iron (III) ions? A. 5 mole B. 4 mole C. 3 mole D. 2 mole E. 1 mole

288) The common feature of reactions at the anode is that A. electrons are consumed B. oxidation is involved C. ions are reduced D. the electrode dissolves E. the electrolyte is diluted

289) The position of equilibrium in a reversible reaction is affected by A. particle size of the reactants B. change in concentration of the reactants C. change in size of the reaction vessel D. vigorous stirring of the reaction mixture E. presence of a catalyst.

290) Separation of mixture of solids by physical methods can be based on differences in the following except A. melting point B. solubility C. particle size D. molar mass E. magnetic property

291) Ethene undergoes mainly addition reactions because it is A. a gas B. a hydrocarbon C. unsaturated D. easily polymerized E. a covalent compound

28

292) The reaction between alkanoic acids and alkanols in the presence of a mineral acid is known as   A. saponification B. hydrolysis C. polymerization D. esterification E. dehydration

293) An elimination reaction occurs when bromoethane is converted to   A.   1,2-dibromoethane B. ethanol C. ethane D. ethyl ethanoate E. ethanoic acid

294) The ash, used for making black soap locally, provides   A. glycerol B. sodium-chloride C. potassium ions D. stearic acid E. glyceride

295) Which of the following is the relative molecular mass of a compound which has empirical formula CH_2O?   (H = 1; C = 12; O = 16)   A. 42   B. 45 C. 126 D. 145 E. 180

296) The by-product of the fermentation of sugar to ethanol is A. propane-1,23-triol   B.   ethyl ethanoate C. ethanedioic acid D. propanol E. carbon (IV) oxide

297) Alkanes react with the halogens mainly by   A. addition B. substitution C. esterification D. oxidation E. combustion

298) The use of diamond in abrasives is due to its   A. high melting point B. durability C. luster D. hardness E. octahedral shape

299) Which of the substances lettered P to V in the equation below would be left on the filter paper if the reaction mixture were filtered?
$P_{(aq)} + Q_{(aq)} \longrightarrow V_{(S)} + R_{(aq)} + T_{(g)}$ A. P   B. Q   C. V   D. T   E. R

300) Which of the following when heated strongly in air will leave a metal as residue?
A. Sodium trioxonitrate(V)   B. Potassium trioxonitrate(V)   C. Silver trioxonitrate (V)   D. Lead (II) trioxonitrate (V)   E. Aluminium trioxonitrate (V)

301) The following oxides react with both acids and bases to form salts except   A. zinc oxide B. lead (II) oxide C. aluminium oxide D. tin (IV) oxide E. carbon (IV) oxide

302) Which of the following metals is not extracted by electrolysis?   A. Iron   B. Sodium   C. Calcium D. Magnesium E. Aluminium

303) Which of the following is an ore of aluminium?   A. Haematite   B. Magnetite C. Siderite D. Bauxite E. Cassiterite

304) Lead and tin are the components of   A. steel B. bronze C. brass D. dentist amalgam E. soft solder

305) An atom with 17 protons, 17 electrons and 18 neutrons has a mass number of   A. 17   B. 18   C. 34   D. 35   E. 52

306) If an element R belongs to the same group as sodium, an aqueous solution of ROH will   A. be neutral B. be acidic C. be coloured D. have pH greater than 7 E. turn blue litmus red

307) Element X, with electronic configuration 2, 8, 2 and element Y with electronic configuration 2, 8, 7 are likely to combine by   A. metallic bonding B. covalent bonding C. electrovalent bonding D. dative bonding E. hydrogen bonding

308) The valence electrons of the element $_{12}Mg$ are in the   A. 1s orbital B. 2s orbital C. $2p_x$ orbital D. $2p_y$ orbital E. 3s orbital

309) What is element Y in the following equation? $^{14}_{7}N + {}^{4}_{2}He -----> {}^{1}_{1}H + Y$
A. Fluorine B. Sodium C. Chlorine D. Oxygen E. Carbon

310) Which of the following Group 1 elements has the highest ionization energy?
A. $_{55}Cs$ B. $_{37}Rb$ C. $_{19}K$ D. $_{11}Na$ E. $_{3}Ll$

311) Which of the following statements is correct about sodium chloride in the solid state?   A. It exists as aggregates of ions   B. It conducts electricity C. its melting point is below 100^0C D. It exists as discrete molecules E. its ions are linked by metallic bonds

312) A mixture of $NaCl_{(s)}$ and $CaCO_{(s)}$ is best separated by   A. dissolution followed by filtration B. sublimation followed by crystallization   C. dissolution followed by evaporation   D. dissolution followed by crystallization E. sublimation followed by dissolution.

313) An aqueous solution is acidic if   A. its pH value is high   B. it is corrosive C. it changes the colour of litmus D. it has a bitter taste E. it contains more H_3O^- than OH^-

314) The following oxides react with water except   A. Na_2O B. SO_3 C. NO_2 D. CO_2 E. CuO

315) Which of the following methods is suitable for the preparation of an insoluble salt? A. Action of an acid on a metal   B. Double decomposition C. Neutralization D. Action of an acid on a trioxocarbonate (IV) E. Action of an acid on an oxide

316) Which of the following will produce oxygen and hydrogen during its electrolysis using platinum electrode? A. Glucose solution B. Aqueous copper (II) tetraoxosulphate (VI)   C. Dilute sodium chloride solution D. Concentrated hydrochloric acid E. Dilute copper (II) chloride solution

317) What mass of copper would be deposited by a current of 1.0 ampere passing for 965 seconds through copper (II) tetraoxosulphate (VI) solution? [Cu = 63.5; 1F = 96500C].   A. 0.318g B. 0.635g C. 3.18g D. 6.35g E. 9.65g

318) How many electrons are removed from Cr^{2-} when it is oxidized to $Cr_2O_7^{2-}$? A. 0 B. 2 C. 4 D. 8 E. 10

319) Substances which absorb water from the atmosphere without dissolving in it are said to be A. anhydrous B. hygroscopic C. deliquescent D. hydrated E. efflorescent

320) Pipe-borne water is usually chlorinated in order to A. improve the taste of the water B. remove the hardness in the water C. coagulate sediments in the water D. kill harmful bacteria in the water E. make the water clear and colourless

321) Rusting of iron is an example of A. deliquescence B. decomposition C. displacement reaction D. redox reaction E. reversible reaction

322) Aluminium is extracted from A. magnetite B. cryolite C. duralumin D. bauxite E. cassiterite

323) Metals are said to be malleable because they A. are good conductors of heat B. are good conductors of electricity C. can be beaten into sheets D. can be alloyed E. can be polished

324) A device used in the laboratory for intermittent production of gases without heat is A. Leibig condenser B. aspirator C. delivery tube D. inverted funnel E. Kipp's apparatus

325) The liquid product of the destructive distillation of coal is A. kerosene B. ethanol C. ammoniacal liquor D. Butane E. Hexane

326) Which of the following can undergo both addition reactions and substitution reactions? A. Benzene B. Pentane C. Propene D. Hexane E. Ethene

327) Petrol consists mainly of A. polymers B. alkanoates C. hydrocarbons D. alkanols E. methanol

328) Dehydration of ethanol produces A. ethanoic acid B. propan-1,2,3-triol C. ethane D. ethanol E. ethene

329) When alkanols react with sodium, the gas evolved is A. hydrogen B. oxygen C. methane D. ethyne E. carbon (IV) oxide

330) A positive reaction to Fehling's test indicates the presence of A. starch B. reducing sugars C. oxidizing agents D. alkanoic acids E. alkanols

331) Which of the following atoms contains the highest number of electrons in the outermost shell? A. $_8O$ B. $_{10}Ne$ C. $_{15}P$ D. $_{19}K$

31

332) Pauli Exclusive Principle is related to   A. quantum number of electrons B. reversibility of equilibrium reactions C. electronegativity values of elements D. collision theory of reaction rates

333) Which of the orbitals 4s, 4p, 4d and 4f has the lowest energy?   A. 4f B. 4p C. 4d D. 4s

334) Which of the following has neither mass nor charge?   A. Alpha particle B. Deuterium C. Gamma ray D. Neutron

335) Two radioactive elements Q and S have half-life periods of 10 hours and 20 hours respectively. Therefore,   A. S decays faster than Q   B. Q is twice as stable as S.   C. Q emits fewer particles than S   D. S is more stable than Q

336) Calcium atom ionizes by   A. gaining two electrons B. losing two electrons C. sharing two electrons D. gaining two protons

337) Which of the following properties increases down a group in the Periodic Table? A. Atomic radius B. Electronegativity C. Electron affinity D. Ionization energy

338) In the Periodic Table, the elements that lose electrons most readily belong to   A. Group I A B. Group IIA   C. Group IIIA   D. Group VIIA

339) Which of the following halogens is the most reactive?   A. F_2 B. Br_2 C. I_2 D. Cl_2

340) The crystal layers in graphite are held together by   A. electrostatic forces B. van der Waals' forces C. hydrogen bonds D. covalent bonds

341) Exhaust fumes discharged from a smoky vehicle gradually become invisible as a result of   A. diffusion B. combustion C. absorption D. emission

342) Which of the following movements can be described as random motion? A. A wheelbarrow being pushed B. A car travelling on a straight line C. Gas molecules colliding in a flask   D. Planets going round the sun

343) Equal volumes of CO_2 cnd CO at s.t.p. have the same   A. mass B. density C. rate of diffusion D. number of molecules

344) Gas molecules are said to be perfectly elastic because   A. they collide without loss of energy B. they move about in straight lines C. the distance between them are negligible D. the volume occupied by them is negligible

345) What volume of hydrogen is produced at s.t.p. when 2.60g of zinc reacts with excess HCl according to the following equation? $Zn_{(s)} + 2HCl_{(aq)} ----->ZnCl_{2(aq)} + H_{2(g)}$ [Zn = 65, 1 mole of a gas occupies 22.4 dm^3 at s.t.p.]
A. 0.040dm^3 B. 0.896dm^3 C. 5.82dm^3 D. 8.62dm^3

346) Helium is preferred to hydrogen in filling balloons because hydrogen A. is inflammable B. is diatomic C. exhibits isotopy D. is a component of water

347) 30cm^3 of hydrogen at s.t.p. combine with 20cm^3 of oxygen to form steam according to the following equation: $2H_{2(g)} + O_{2(g)} -----> 2H_2O_{(g)}$. calculate the total volume of the gaseous mixture at the end of the reaction A. 50cm^3 B. 35cm^3 C. 30cm^3 D. 25cm^3

348) An exothermic reaction is one which involves A. attainment of dynamic equilibrium B. loss of heat to the surrounding C. evolution of gas as it proceeds D. positive change in value of enthalpy

349) Which of the following is an acid salt? A. NH_4Cl B. $MgSO_4 . 7H_2O$ C. CH_3COONa D. $NaHCO_3$

350) Which of the following compounds dissolves in water to form a solution with pH below 7? A. Sodium tetraoxosulphate (VI) B. Potassium hydroxide C. Ammonium chloride D. Sodium trioxocarbonate (IV)

351) If the solubility of KHCO$_3$ is 0.40 mol dm^{-3} at room temperature, calculate the mass of KHCO$_3$ in 100cm^3 of the solution at this temperature.[KHCO$_3$ = 100gmol^{-1}] A. 4.0g B. 10.0g C. 40.0g D. 100.0g

352) The property of calcium chloride which makes it useful as a drying agent is that it is A. ionic B. deliquescent C. a strong electrolyte D. a normal salt

353) Which of the following lowers the activation energy of a chemical reaction? A. Freezing mixture B. Reducing agent C. Water D. Catalyst

354) Which of the following is not correct about oxidation number convention? The oxidation number of a/an A. element in the free state is zero B. monatomic ion is equal to its charge C. alkali metal like sodium in its compounds is +1 D. atom of a diatomic molecule is +2

355) Equal volumes of water were added to one mole of each of the following compounds. Which of them produced the largest number of ions. A. Glucose B. Sodium trioxonitrate (V) C. Ethanoic acid D. Silver chloride

356) A metal which can be used as sacrificial anode for preventing corrosion of a length of iron pipe is   A. copper B. magnesium C. silver D. lead

357) Separating funnel is used for separating a mixture of   A. liquids with different boiling points   B. sediments in a liquid   C. liquids with different colours D. liquids that are immiscible

358) How many carbon atom are present in one molecule of 2-methylpropane?   A. 2 B. 3 C. 4 D. 5

359) Which of the following is the most reactive towards bromine? A. Methane B. Benzene C. Ethyne D. Hexane

360) Which of the following compounds is unsaturated? A. 2-methylbutane B. Chloromethane C. Methylbenzene D. 1, 2-dibromoethane

361) Geometric (cis-trans) isomerism is exhibited by   A. $C_2H_2Cl_2$ B. C_2H_5Cl C. C_4H_{10} D. C_5H_{12}

362) Which of the following is needed to produce slag during the extraction of iron in the blast furnace?   A. Cryolite B. Limestone C. Carbon D. Sulphur

363) Environmental pollution problems are generally more severe in countries which have   A. dense population B. low birth rate C. great land mass D. limited industries

364) The greenhouse effect is a climatic condition associated with the presence of excess   A. carbon (IV) oxide B. hydrogen sulphide C. Nitrogen (II) oxide D. ammonia gas

365) The atoms of four elements are represented as $_{20}Q$, $_{16}R$, $_{10}S$ and $_{8}T$. Which of the elements would be unreactive? A. Q   B. R.   C. S   D. T

366) Chlorine atom forms Cl⁻ by   A. losing one electron   B. sharing one electron C. donating lone pair of electrons   D. gaining one electron

367) Electrovalent compounds are characterized by   A. solubility in ethanol   B. high molar mass C. high melting point D. strong oxidizing ability

368) The type of chemical bond that exists between potassium and oxygen in potassium oxide is   A. ionic B. metallic C. covalent   D. dative

369) Which of the following contains co-ordinate covalent bond?   A. HCl B. CH_4 C. NH_4^- D. NaCl

370) Electrostatic force of attraction between sodium ion and halide ion is greatest in   A. NaCl   B. NaBr   C. NaF   D. NaI

371) Which of the following non-metals reacts most readily with metals?   A. Nitrogen   B. Chlorine   C. Sulphur   D. Carbon

372) A radioactive solid is best stored   A. under paraffin oil   B. under ultraviolet light   C. in a cool, dark cupboard   D. in a box lined with lead

373) What is the shape of a molecule of CCl_4?   A. Pyramidal   B. Tetrahedral   C. Trigonal planar   D. Linear

374) Which of the following increases as boiling water changes to steam?   A. Temperature of the system   B. Degree of disorder in the system   C. Number of molecules   D. Activation energy

375) 9.60g of a gas X occupies the same volume at 0.30g of hydrogen under the same conditions. Calculate the molar mass of X (H = 1).   A. $8g\ mol^{-1}$ B. $16g\ mol^{-1}$ C. $32g\ mol^{-1}$ D. $64g\ mol^{-1}$

376) Which of the following gases is colourless, odourless and soluble in potassium hydroxide solution?   A. NO_2 B. SO_3 C. NH_3 D. CO_2

377) The gas evolved when dilute tetraoxosulphate (VI) acid reacts with sodium hydrogen trioxocarbonate (IV) is   A. hydrogen   B. oxygen   C. carbon (IV) oxide   D. sulphur (VI) oxide

378) If a solution has a pH of 2, it can be concluded that it   A. is a weak electrolyte   B. has hydrogen ion concentration of $0.2\ mol\ dm^{-3}$ C. Is twice as acidic as a solution of pH 1.   D. will produce effervescence with magnesium ribbon

379) Which of the following compounds dissolves in water to give an alkaline solution?   A. Ammonium chloride   B. Sodium chloride   C. Sulphur (IV) oxide   D. Potassium oxide

380) Which of the following salts is stable to heat?   A. K_2CO_3 B. $(NH_4)_2SO_4$ C. $NaHCO_3$ D. $AgNO_3$

381) What type of reaction is involved when wood shavings produce a brown gas with concentrated HNO_3?   A. Redox reaction   B. Destructive distillation   C. Dehydration   D. Neutralization

382) Chemical equilibrium is attained when   A. reactants in the system are used up   B. concentrations of the products are greater than those of the reactants   C. concentrations of the reactants and products remain constant   D. reactants stop forming the products

383) A finely divided form of a metal burns more readily in air than the rod form because the rod has A. higher molar mass B. smaller surface area C. protective oxide coating D. different chemical properties

384) Which of the following will not displace copper from a solution of copper (II) salt? A. Aluminium ions B. Magnesium C. Silver D. Zinc ions

385) In which of the following is the oxidation number of sulphur equal to -2?
A. S_8 B. H_2S C. SO_2 D. SO_3^{2-}

386) 0.10 mol dm^{-3} NaCl conducts electricity better than 0.10 mol dm^{-3} CH_3COOH because the solution of NaCl A. contains more ions B. is neutral to litmus C. has a lower molar mass D. has a higher pH.

387) Which of the following procedures is suitable for identifying the food colouring added to a pastry sample? A. Carry out some food tests B. Examine the pastry sample under a microscope C. Do a chromatogram of the colouring extract D. Determine the molar mass of the colouring

388) C_8H_{18} will undergo the following reactions except A. cracking B. combustion C. substitution D. addition

389) Which of the following is correct about both compounds shown below? $CH_3CH_2CH_2OH$; $CH_3CH_2CH(OH)CH_3$ A. They have identical physical properties B. They are structural isomers C. Their general molecular formula is the same D. They are oxidized to alkanones

390) Which of the following needs to be hydrolysed before it shows reducing property? A. Glucose B. Maltose C. Sucrose D. Fructose

391) Two-organic substances are labelled K and L. If K gave a blue-black colour with iodine solution and L gave a deep red precipitate with Million's reagent, it can be concluded that A. K is a carbohydrate and L is a protein B. K is alkaline and L is acidic C. K is unsaturated and L is a reducing sugar D. K is a reducing agent and L is an amino acid

392) A colourless, odourless liquid T, gives effervescence with sodium trioxocarbonate (IV) and a white precipitate with silver trioxonitrate (V) solution. T is more probably A. sodium chloride solution B. barium chloride solution C. dilute trioxonitrate (V) acid D. dilute hydrochloric acid

393) When sodium hydroxide solution is added to a solution of zinc salt, the white precipitate formed redissolves in excess sodium hydroxide because A. sodium is more reactive than

zinc   B. sodium hydroxide is a strong alkali   C. zinc hydroxide is amphoteric   D. zinc hydroxide is unstable

394) In the electrolytic extraction of aluminium from purified alumina, molten cryolite is added in order to   A. lower the melting point of the alumina B. prevent the aerial oxidation of the molten aluminium   C. lower the activation energy of the reaction   D. form a protective crust on top of the electrolyte

395) If $20cm^3$ of distilled water is added to 80 cm^3 of 0.50 mol dm^{-3} hydrocloric acid, the concentration of the acid will change to   A. 20 mol dm^{-3} B. 0.40 mol dm^{-3} C. 2.00 mol dm^{-3} D. 5.00 mol dm^{-3}

396) Which of the following pollutants encourage the growth of algae in rivers?   A. Oil spills   B. Phosphates from detergents   C. Suspended solids D. Pesticide residues

397) Biotechnology is applied in the following **except**   A. baking industry B. wine production C. production of antibiotics   D. manufacture of soap

398) A plastic which cannot be softened by heat is described as   A. thermosetting   B. non-biodegradable   C. thermoplastic D. malleable

399) Allotropes of an element differ in their   A. physical properties B. chemical properties C. mass numbers D. electronic configuration

400) What is the mass number of an element if its atom contains 10 protons, 10 electrons and 12 neutrons?   A. 32   B. 22   C. 20   D. 10

401) Given that the electronic configuration of an element X is $1s^2 2s^2 2p^6 3s^2 3p^4$, it can be deduced than X A. belongs to group VI in the Periodic Table   B. belongs to period 4 in the Periodic Table   C. contains 3 unpaired electrons in the ground state   D. has atomic number 27

402) P, Q, R and S are metals in the same group in the Periodic Table but in periods 3, 4, 5 and 6 respectively. Which of them loses electrons least readily?   A. P   B. Q   C. R   D. S

403) Elements which belong to the same group in the Periodic Table are characterized by   A. difference of +1 in the oxidation numbers of successive members   B. presence of the same number of outermost electrons in the respective atoms   C. difference of 14 atomic mass units between successive members   D. presence of the same number of electron shells in the respective atoms

404) The atom of an element X has two electrons in its outermost shell. What is the formula of the compound formed when X combines with aluminium ($_{13}Al$)?   A. AlX_2 B. Al_2X C. Al_2X_2 D. Al_2X_3

405) The presence of unpaired electrons in an atom of a transition metal gives rise to   A. paramagnetism   B. malleability   C. ductility   D. shiny appearance

406) What is responsible for metallic bonding?   A. Sharing of electrons between the metal atoms   B. Attraction between the atomic nuclei and cloud of electrons   C. Transfer of electrons from one atom to another   D. Attraction between positive and negative ions

407) Which of the following compounds is covalent?   A. $CaCl_2$ B. MgO C. NaH D. CH_4

408) A sheet of paper is placed in the path of a beam from a radioactive source. The emissions that pass through the paper consist of   A. alpha particles and gamma rays   B. alpha and beta particles   C. beta particles and gamma rays   D. alpha particles only

409) Which of the following features of a human skeleton can be determined by carbon-dating?
        A. Height   B. Mass   C. Age   D. Race

410) $P_{Total} = P_1 + P_2 + P_3 +\ldots\ldots\ldots.P_n$, where P_{Total} is the pressure of a mixture of gases. The equation above is an expression of   A. Graham's law   B. Gay-Lussac's law   C. Boyle's law
        D. Dalton's law

411) Which of the following pairs of reagents react to produce hydrogen?   A.   Zinc   and concentrated trioxonitrate (V) acid B. Water and calcium dicarbide   C.   Copper and dilute hydrochloric acid   D. Magnesium and dilute tetraoxosulphate (VI) acid

412) A mixture of calcium chloride and calcium trioxocarbonate (IV) in water can be separated by
        A. evaporation   B. sublimation   C. distillation   D. Filtration

413) Which of the following would dissolve in warm dilute H_2SO_4 without effervescence, to give a blue solution? A. Copper turnings   B. Copper (II) trioxocarbonate (IV)   C. Copper (II) Oxide   D. Copper alloy

414) The gas evolved when dilute hydrochloric acid is added to limestone is   A. NH_3 B. SO_2
        C. HCl   D. CO_2

415) An aqueous solution of sodium trioxocarbonate (IV) is   A. organic   B.   acidic C. alkaline   D. neutral.

416) What is the oxidation number of nitrogen in $Al(NO_3)_3$?   A.   +1   B. +3   C. +5
        D. +6

417) Which of the following substances decomposes when an electric current is passed through it?
        A. Glucose solution   B. Zinc rod   C. Hydrochloric acid   D. Tetrachloromethane

418) A concentrated solution containing H^+, Cu^{2+}, OH^- and Cl^- was electrolysed using platinum electrodes. Which of the ions will be discharged at the cathode?
A. H^+ B. Cu^{2+} C. OH^- D. Cl^-

419) Given that the order of reactivity of four metals is P>Q>R>S, which of the following reactions is feasible?   A. $S + P^+ -----> P + S^+$ B. $Q + S^+ -----> S + Q^+$ C. $R + P^+ -----> P + R^+$ D. $S + Q^+ -----> Q + S^+$

420) In the electrolysis of $CuSO_{4(aq)}$ using copper electrodes, the reaction at the anode is   A. discharge of SO_4^{2-} B. discharge of OH^- C. dissolution of copper electrode   D. evolution of oxygen

421) Hydrogenation of butene yields   A. butyne   B. butane   C. pentene   D. butanol

422) Benzene produces more soot than ethane on burning because benzene   A.   has a higher molar mass   B. has a higher degree of unsaturation   C. undergoes both substitution and addition reactions   D. is a liquid at room temperature

423) One of the products of combustion of pentane in excess air is   A. pentanol   B. pentene   C. nitrogen (II) oxide D. carbon (IV) oxide

424) Which of the following polymers contains nitrogen?   A. Nylon   B. Polyvinyl chloride   C. Polyethene   D. Cellulose

425) Which of the following establishments uses the process of fermentation in its operation?
        A. Fertilizer plant   B. Textile industry   C. Brewery   D. Soap manufacturing industry

426) Sugars can be hydrolyzed at room temperature by the action of   A. mineral acids   B. strong bases C. enzymes D. Oxidants

427) Which of the following constitutes an advantage in the use of hard water?   A. Formation of scum with soap   B. Furring of kettles   C. Blockage of water pipes   D. Formation of strong bones

428) Which of the following compounds is readily soluble in water?   A. CuO B. AgCl   C. Na_2SO_4 D. $CaCO_3$

429) If $10cm^3$ of distilled water is added to $10cm^3$ of an aqueous salt solution, the concentration of the solution   A. increases B. Decreases C. remains constant   D. doubles

430) The main type of reaction that occurs in the blast furnace during the extraction of iron is   A. reduction reaction B. decomposition   C. exothermic reaction D. combustion

431) Acidic industrial waste can be treated with   A. lime   B. brine   C. water   D. enthanol

432) Waste plastics accumulate in the soil and pollute the environment because plastic materials are   A. insoluble in water   B. non-biodegradable   C. easily affected by heat   D. inflammable

433) If the relative molecular mass of an element is not a whole number, it can be deduced that the element is   A. naturally radioactive   B. abundant in nature   C. a transition metal   D. an isotopic mixture

434) The atom of an element X is represented as ^Y_ZX. The basic chemical properties of X depend on the value of   A. Y   B. Z   C. Y – Z   D. Z – Y

435) The atomic number of chlorine is 17. What is the number of electrons in a chloride ion? A. 16   B. 17   C. 18   D. 19

436) A hydrogen atom which has lost an electron contains   A. one proton only   B. one neutron only   C. one proton and one neutron   D. one proton, one electron and one neutron

437) Which of the following elements has a ground state electronic configuration with an incomplete penultimate shell?   A. Iron   B. Sodium   C. Aluminium   D. Calcium

438) An element Y has the electronic configuration $1s^2 2s^2 2p^6 3s^2 3p^4$. To what period does it belong in periodic table?   A. 3   B. 4   C. 5   D. 6

439) Which of the following decreases when a given mass of gas is compressed to half its initial volume?   A. Average intermolecular distance   B. Frequency of collisions   C. Number of molecules present   D. Atomic radius of each particle

440) The gas given off when NH_4Cl is heated with an alkali is   A. H_2 B. Cl_2 C. N_2 D. NH_3

441) A major factor considered in selecting a suitable method for preparing a simple salt is its A. crystalline form   B. melting point   C. reactivity with dilute acids   D. solubility in water

442) Which of the following acids forms normal salts only?   A. Tetraoxosulphate (VI) acid B. Trioxosulphate (IV) acid   C. Tetraoxophosphate (V) acid   D. Trioxonitrate (V) acid

443) A substance is said to be hygroscopic if it absorbs   A. water from the atmosphere to form a solution   B. heat from the surrounding   C. carbon (IV) oxide from the atmosphere   D. moisture from the atmosphere without dissolving

444) Which of the following substances increases in mass when heated in air?
A. Sodium chloride   B. Iodine crystals   C. Magnesium ribbon D. Copper (II) oxide

445) Calculate the mass of sodium hydroxide in 5.00 dm^3 of a 0.125 mol dm^{-3} solution. [NaOH = 40 g mol^{-1}]   A. 0.0156g   B. 0.625 g   C. 1.00 g D. 25.0 g

446) Which of the following salt solutions will have a pH greater than 7?
A. $NaCl_{(aq)}$ B. $Na_2CO_{3(aq)}$ C. $Na_2SO_{4(aq)}$ D. $NaHSO_{4(aq)}$

447) Equal amounts of marble chips are reacted separately with 100cm^3 of hydrochloric acid of different concentrations. If all the marble chips reacted, which of the following remained the same in each case?   A. Average rate of evolution of gas B.   Total mass of gas evolved during the reaction   C. Time taken for the reaction to reach completion   D. Initial reaction rates

448) $CuSO_4.5H_2O$ can be obtained from an aqueous solution to copper (II) tetraoxosulphate (VI) by
        A. evaporation to dryness   B. using chromatography   C. precipitation   D. crystallization

449) Which of the following compounds will leave a metal residue when heated?   A. $Cu(NO_3)_2$
B. $AgNO_3$ C. K_2CO_3 D. $CaCO_3$

450) A visible change is observed when a strip of iron is placed in an aqueous solution of   A. $FeSO_4$
B. $ZnSO_4$ C. $CuSO_4$ D. $MgSO_4$

451) Which of the following ions require the largest quantity of electricity for discharge at an electrode?   A. 2.0 mole of Q^{3+} B. 2.5 mole of R^{2+} C. 3.0 mole of T^-
D. 4.0 mole of Y^-

452) If two metallic ions of the same concentration migrate to a graphite cathode, the one that would be preferentially discharged is the one that has the   A. smaller mass B lower electrical charge C. greater stability in solution D. stronger tendency to accept electrons

453) When concentrated sodium chloride solution is electrolysed using inert electrodes, the products are   A. oxygen and hydrogen B. hydrogen and chlorine C. sodium and oxygen D. sodium and chlorine

454) Which of the following processes takes place during the production of margarine from vegetable oils?   A. Esterification B. Hydrolysis C. Hydrogenation D. Saponification

455) Which of the following compounds can be represented by the molecular formula C_2H_6O?   A. Propanal   B. Ethanol   C. Methanoic acid D. Glucose

456) When alkynes are hydrogenated completely, they produce compounds with the general molecular formula   A. C_nH_n B. C_nH_{2n+2} C. C_nH_{2n} D. C_nH_{2n-2}

457) The following compounds are hydrocarbons except   A. methylpropanoate   B. 2-methylbutane C. Benzene D. cyclohexane

458) The reaction between ethane and chlorine to form 1, 2-dichloroethane is
A. oxidation B. polymerization C. addition D. substitution

459) Which of the following types of alkanols undergo oxidation to produce alkanoic acids?   I. Primary alkanols II. Secondary alkanols III. Tertiary alkanols
A. I, II and III   B. I and II only   C. II only   D. I only

460) A suitable reagent for distinguishing between ethanoic acid and ethanol is   A. bromine water B. Fehling's solution C. sodium hydrogentrioxocarbonate (IV)   D. ammoniacal silver trioxonitrate (V)

461) The reaction of sucrose with dilute HCl produces   A. an alkanoate   B. glucose and fructose
C. a black mass of carbon D. a polysaccharide

462) One of the methods used for combating air pollution is   A. burning of solid waste   B. provision of sanitary land fills C. legislation against some industrial practices   D. scavenging of refuse dumps

463) The electronic configuration of an element is $1s^2 2s^2 2p^6 3s^2 3p^3$. Where is the element located in the periodic table?   A. Group III, period 3 B. Group III, period 5 C. Group V, period 3 D. Group V, period 5

464) The electrons that are most easily removed from $^{24}_{12}Mg$ are in which orbital?   A. 1s B. 2s C. 2p D. 3s

465) Group 0 elements are unreactive because   A. their outermost shells contain maximum number of electrons B. they occupied the highest energy level   C. they are monatomic gases D. they are non-metals

466) An element X with electronic configuration $1s^2 2s^2 2p^6 3s^2$ combines with another element Y with the configuration $1s^2 2s^2 2p^6 3s^2 3p^5$. What is the formula of the compound formed? A. X_2Y_5 B. X_2Y_3 C. XY_3 D. XY_2

467) Which of the following statements about dative bonding is not correct?   A. There is sharing of electrons B. One atom must possess a lone pair of electrons   C. Each participating atom contributes one electron D. It can be formed between molecules

468) If 11g of a gas occupies 5.6 dm^3 at s.t.p., calculate its vapour density. (1.0 mole of a gas occupies 22.4dm^3 at s.t.p.   A. 22   B. 44   C. 66   D. 88

469) Which of the following acids is dibasic?   A. Hydrochloric acid   B. Trioxosulphate (IV) acid   C. Dioxonitrate (III) acid   D. Ethanoic acid

470) Which of the following salts can be prepared by reacting the metal with dilute H_2SO_4?   A. $CuSO_4$ B. $MgSO_4$ C. $BaSO_4$ D. $PbSO_4$

471) Which of the following solutions will give a precipitate with dilute tetraoxosulphate (VI) acid?   I. Lead (II) trioxonitrate (V) solution   II. Barium chloride solution   III. Sodium chloride solution   A. I only   B. II only   C. I and II only   D. II and III only

472) Substance Q reacts with cold water to form a solution which turns red litmus blue. Substances Q could be   A. zinc   B. phosphorus   C. lead   D. sodium

473) What term is used to describe an oxide whose aqueous solution turns blue litmus red?   A. Strong electrolyte   B. Acid anhydride   C. Amphoteric oxide   D. Basic oxide

474) Which of the following processes involves neutralization?   A. Hardening of oils   B. Souring of milk   C. Charring of sugar   D. Liming of soil

475) Sea shells contain mainly $CaCO_3$. Calcium oxide can be prepared from sea shells by   A. adding dilute acid   B. heating at a high temperature   C. reduction with CO   D. adding dilute $NaOH_{aq)}$

476) Which of the following statements about an exothermic reaction is correct?   A. The products have less heat content than the reactants   B. The system absorbs heat from the surrounding   C. The activation energy is high   D. The enthalpy change is positive

477) A reducing agent is expected to   A. decolorize acidified $KMnO_4$ solution   B. decolorize acidified $FeSO_4$ solution   C. Liberate Cl_2 from a chloride   D. liberate CO_2 from $NaHCO_3$

478) Five metals represented by the letters V to Z are arranged in order of their reactivity as shown below V>W>X>Y>Z
Which of the metals cannot displace Y from an aqueous solution of its salt?
A. V   B. W   C. X   D. Z

479) Which of the metals is most likely extracted by the electrolysis of its fused chloride?   A.   V   B. W   C. X   D. Y

480) Which of the following statements is/are correct about tin plating?   I. The electrolyte must be a soluble salt of tin   II. The object to be plated is made the cathode   III. The anode is made of tin   IV. The negative electrode dissolves   A. I only   B. I and II only   C. I, II and III only   D. I, II and IV only

481) Which of the following compounds has the same empirical formula as benzene? A. C_2H_2 B. C_2H_4 C. C_4H_6 D. C_6H_{12}

482) A given fuel has an octane number of 100. This means that A. the knocking tendency of the fuel is low B. the fuel is a mixture of several hydrocarbons C. a large amount of alkanes is present in the fuel D. the fuel needs to be cracked before use

483) Wine containing 8% to 17% ethanol can be converted to gin containing about 40% ethanol by A. evaporation B. distillation C. fermentation D. oxidation

484) The reaction of vegetables oil with a solution of wood ash is A. saponification B. neutralization C. hydrogenation D. esterification

485) What process does the following equation represent? $(C_6H_{10}O_5)_n + nH_2O \longrightarrow nC_6H_{12}O_6$? A. Polymerization of glucose B. Hydrolysis of starch C. Fermentation of sugar D. Dehydration of carbohydrates

486) When excess ethene is shaken with acidified $KMnO_4$ solution, the product obtained is A. ethane B. ethanol C. ethane - 1,2-diol D. ethanoic acid

487) What volume of 0.20 mol dm^{-3} NaOH solution would yield 5.0g of NaOH on evaporation to dryness? [NaOH = 40g mol^{-1}] A. 400cm^3 B. 625cm^3 C. 1000cm^3 D. 1600cm^3

488) What is the main impurity in haematite? A. $CaSiO_3$ B. $CaCO_3$ C. SiO_2 D. Fe_2O_3

489) When a sample of water was boiled, it lathered more readily with soap. It can be concluded that the sample most likely contained A. magnesium tetroxosulphate (VI) B. suspended solids C. Organic impurities D. calcium hydrogen-trioxocarboante (IV).

490) What are the two gases associated with the formation of acid rain? A. CO_2 and HCl B. CO_2 and N_2 C. CO_2 and NO_2 D. HCl and SO_2

491) Which of the following processes will pollute water? A. Exposure of a body of water to ultraviolet rays B. Discharge of industrial effluents into waterways C. Passage of river water through a sand bed D. Addition of a measured quantity of chlorine to water

492) The mass of an element is 27 and its atomic number is 13. What is the composition of the nucleus of its atom? A. 13 electrons and 14 protons B. 13 neutrons and 14 protons C. 13 protons and 14 neutrons D. 13 electrons and 14 neutrons

493) If an atom is represented as $^{39}_{19}X$, which of the following deductions is correct? A. It contains 20 protons B. It forms a covalent chloride C. its atomic number is 39 D. It is an alkali metal

494) Two elements X and Y combine to form a compound with formula X_2Y_3. Which of the following representations would fit the configurations of X and Y?

	X	Y
A.	2, 2	2, 3
B.	2, 6	2, 5
C.	2, 8, 2	2, 8, 3
D.	2, 8, 3	2, 6

495) Which of the electrons in the following orbitals will experience the greatest nuclear attraction? A. 1s electron of helium B. 1s electron of potassium C. 2s electron of boron D. 2s electron of carbon

496) A nuclide emits a beta-particle (e). In the process, its atomic number A. increases by 1 while the mass number remains constant B. decreases by 2 while the mass number remains constant C. remains constant while the mass number decreases by 2. D. remains constant while the mass number increases by 1

497) One of the characteristics of transition metals is A. reducing ability B. ductility C. ability to conduct electricity D. formation of coloured ions

498) How many electrons are in the L shell of $^{31}_{15}P$? A. 2 B. 5 C. 8 D. 16

499) A metal X forms two chlorides with the formulae XCl_2 and XCl_3. Where is X in the Periodic Table? A. Group II B. Group III C. d-block D. s-block

500) What type of bonding exists between X and chlorine in each of the chlorides above? A. metallic bonding B. ionic bonding C. covalent bonding D. dative bonding

501) Which of the following equipment is not used to detect radioactivity? A. Wilson cloud chamber B. Geiger-Muller counter C. Mass spectrometer D. Photographic plate

502) The method of collection of gases prepared in the laboratory depend on their A. odour and atomicity B. colour and odour C. atomicity and density D. solubility and density

503) The alloy used extensively in the building industry is A. steel B. bronze C. duralumin D. type metal

45

504) Which of the following salts is not prepared by precipitation?   A.   Lead   (II) trioxocarbonate(IV)   B. Barium trioxocarbonate (IV)   C. Sodium trioxocarbonate (IV)   D. Calcium trioxocarbonate (IV)

505) What volume of hydrogen would be left over when $300cm^3$ of oxygen and $1000cm^3$ of hydrogen are exploded   A. $400cm^3$ B. $700cm^3$ A. $650cm^3$ B. $350cm^3$

506) Which of the following substances will not produce effervescence with dilute hydrochloric acid?   A. Copper (II) trioxocarbonate (IV)   B. Potassium hydrogen-trioxocarbonate (IV)   C. Zinc granules   D. Sodium chloride

507) What quantity of electrons in moles is needed to discharge two moles of aluminium from aluminium oxide (Al_2O_3)?   A. 1   B. 2   C. 4   D. 6

508) A solution of zinc chloride should not be stored in a container made of   A. tin   B. copper   C. aluminium   D. lead

509) Glucose gives a brick-red precipitate with Fehling's solution because it is   A.   a carbohydrate   B. an alkanoate   C. a reducing sugar   D. a non-electrolyte

510) What is the IUPAC name of $CH_3CH_2COOCH_3$?   A. Ethylethanoate   B. Methylethanoate   C. Methylpropanoate   D. Propylmethanoate

511) The mass of $800cm^3$ of a gas X at s.t.p. is 1.0g. What is the molar mass of X? [1 mole of a gas at s.t.p. occupies $22.4dm^3$]   A. $18.0g\ mol^{-1}$ B. $22.4g\ mol^{-1}$ C. $28.0g\ mol^{-1}$ D. $36.0g\ mol^{-1}$

512) Which of the following pairs of acid and base will produce a solution with pH less than 7 at equivalent point? A. HNO_3 and $NaOH$ B. H_2SO_4 and KOH C. HCl and $Mg(OH)_2$ D. HNO_3 and KOH

513) Aqueous copper (II) tetraoxosulphate (VI) was electrolysed using the following pairs of electrodes, in which case was there a decrease in the mass of the anode?

Cathode	Anode
A. Graphite	Graphite
B. Platinum	Platinum
C. Copper	Copper
D. Graphite	Platinum

514) Which of the following salts is readily soluble in water?   A. $PbCl_2$ B. $Pb(NO_3)_2$ C. $PbSO_4$ D. $PbCO_3$

515) An orange drink concentrate is suspected to contain traces of poisonous green dye and a harmless dye having the same boiling point. Which of the following techniques is most

46

suitable for isolating the dyes?　　A. Fractional distillation　B. Paper chromatography　C. Coagulation　D. Recrystallization

516) When ammonium trioxocarbonate (IV) solution is added to separate solutions of calcium, zinc and sodium salt, precipitate could be formed in the case of
A. calcium only　B. zinc only　C. calcium and zinc only　D. sodium and zinc only

517) Consider the following electronegative values
H,　　W,　　X,　　Y,　　Z
2.1　　1.2　　1.5　　2.1　　3.5
Which of the elements represented by the letters W to Z will combine with hydrogen (H) to form a purely covalent compound?　　A. W　B. X　C. Y　D. Z

518) The condensation of several amino acid molecules gives　　A. long chain alkanoic acids　B. secondary alkanols　C. alkanoates　D. proteins

519) The key factor to be considered in siting a chemical industry is　　A. favourable climatic conditions　B. the availability of space to store raw materials　C. Its nearness to other industrial establishments　D. its nearness to the source of raw materials

520) Which of the following petroleum fractions yields Vaseline and paraffin wax on redistillation?
A. Diesel oil　B. Lubrication oil　C. Petrol　D. Kerosene

521) What is the product CxHy in the following equation?　$C_{10}H_{22} \longrightarrow C_8H_{18} + CxHy$　A. butane B. octane　C. ethene　D. decane

522) Which of the following pollutants is biodegradable?　　A. Domestic sewage　B. metal scraps　C. Radioactive waste　D. Plastic foil

523) The process of 'seeding' during the extraction of aluminium involves the addition of pure $Al(OH)_3$ crystals in order to　A. lower the temperature　B. increase the concentration　C. facilitate the precipitation of solid　D. reduce the viscosity of the solution

524) Which of the following measures is not suitable for controlling water pollution?　A. Treating industrial effluent and domestic sewage before discharge into water　B. Controlling the use of agrochemicals　　C. Recycling industrial and agricultural wastes　D. Burning industrial and agricultural wastes before discharge into water ways.

525) The electronic configuration of an element X is $1s^2 2p^2 2p^6 3s^2 3p^4$. It can be deduced that X
A. belongs to group 6 of the Periodic Table　B. belongs to period IV of the Periodic Table　C. has 3 unpaired electrons in its atom　D. has relative molecular mass of 16.

526) Which of the following is arranged in order of decreasing atomic size in the Periodic Table? A. $_6C, _4Be, _7N, _8O$ B. $_4Be, _6C, _7N, _8O$ C. $_8O, _7N, _6C, _4Be$ D. $_7N, _8O, _6C, _4Be$

527) Which of the following changes is characteristics of an alpha-particle emission? A. The nucleus loses two protons and two neutrons B. A proton combines with an electron giving off heat C. The nucleus splits into two equal parts D. Four neutrons are lost

528) An element X has an atomic number of 16. What is its most likely oxidation state in its binary compounds? A. -3 B. -2 C. +2 D. +4

529) An element which can exist in two or more forms in the same physical state, exhibits A. isotopy B. structural isomerism C. allotropy D. variable valency

530) Which of the following can be used to predict the type of bonding in HCl? A. pH value B. Electronegativity difference C. Heat of neutralization D. Heat of solution

531) Which of the following molecules has a linear structure? A. NH_3 B. H_2O C. CO_2 D. CH_4

532) The volume occupied by a given mass of gas depends on its A. diffusion rate B. temperature and pressure C. degree of solubility D. relative density

533) The activation energy of a reaction can be altered by A. adding a reducing agent B. applying a high pressure C. using a catalyst D. changing the temperature

534) Which of the following factors does not affect the rate of reaction of $CaCO_3$ with HCl? A. Temperature of the reaction B. Solubility of the $CaCO_3$ C. Concentration of the HCl D. Surface area of the $CaCO_3$

535) When X is heated with manganese (IV) oxide, oxygen is produced. X is A. KCl B. $KClO_3$ C. $CaCO_3$ D. $ZnCO_3$

536) Which of the following statements is not a postulate of the kinetic theory of gases? A. Molecules move with the same speed B. Intermolecular forces are negligible C. Molecules are in a state of constant random motion D. Collision between molecules is elastic

537) HNO_3 does not usually react with metals to liberate hydrogen because HNO_3 is A. a strong acid B. a corrosive acid C. an oxidizing agent D. a nitrating agent

538) Which of the following is not an acid anhydride? A. P_2O_5 B. NO_2 C. SO_2 D. CO

539) $NaHCO_{3(S)}$ can be distinguished from $Na_2CO_{3(S)}$ by the action of A. dilute hydrochloric acid B. carbon (IV) oxide C. aqueous ammonia D. heat

48

540) What process is involved in the reaction represented by the following equation? A. $AlCl_3$ + $3H_2O_{(l)}$ -----> $Al(OH)_{3(s)}$ + $3HCl_{(g)}$ A. Dehydration B. Hydrolysis C. Double decomposition D. Neutralization

541) A salt that absorbs moisture from the atmosphere without forming a solution is said to be
A. efflorescent B. deliquescent C. hygroscopic D. insoluble

542) Which of the following compounds dissolves readily in water?   A. $BaSO_4$ B. $CuCO_3$
C. NH_4Cl D. $AgCl$

543) When ice is changing to water, its temperature remains the same because the heat gained is
A. used to separate the molecules B. lost partially to the atmosphere C. used to increase the volume of ice   D. less than the activation energy

544) Which of the following methods is suitable for preparing insoluble salts?   A. Thermal decomposition B. Oxidation C. Double decomposition D. Neutralization

545) A solution of sodium hydroxide containing 6.0g in $250cm^3$ of solution has a concentration of
A. 0.04 mol dm^{-3} B. 0.60 mol dm^{-3} C. 0.96 mol dm^{-3} D. 0.15 mol dm^{-3} [molar mass of NaOH = 40g mol^{-1}]

546) Which of the following elements will burn in excess oxygen to form a product that is neutral to litmus?   A. Carbon B. Hydrogen C. Sulphur D. Sodium

547) Determine the mass of copper deposited by 4.0 moles of electrons in the reaction represented by the equation below: $Cu^{2+}_{(aq)}$ + 2e -----> $Cu_{(s)}$ A. 32 B. 64 C. 128 D. 256

548) The main difference between a primary cell and a secondary cell is that the primary cell A. is an electrolytic cell but the secondary is not B. cannot be recharged but the secondary cell can   C. can be recharged but the secondary cell cannot D. contains electrodes but the secondary cell does not

549) A solid can be obtained from its saturated solution by   A. filteration   B. crystallization C. condensation D. decomposition

550) What is the reducing agent in the reaction represented by the following equation?
$Fe^{3+}_{(aq)}$ + $H_2S_{(g)}$ -----> $2Fe^{2+}_{(aq)}$ + $2H^+_{(aq)}$ + $S_{(s)}$ A. $Fe^{3+}_{(aq)}$ B. $H_2S_{(g)}$ C. $Fe^{2+}_{(aq)}$ D. $S_{(s)}$

551) Reactions undergone by compounds with the general formula C_nH_{2n+2} include
A. addition B. esterification C. substitution   D. dehydration

552) What is X in the reaction represented by the equation below? $C_{10}H_{22} -----> C_2H_6 + C_2H_4 + 3X$
A. Propyne B. Propanol C. Propene D. Propane

553) Which of the following substances is a non-reducing sugar? A. Sucrose B. Glucose C. Fructose D. Maltose

554) The colour of the solution formed when ethyne reacts with an acidic solution of potassium tetraoxomanganate (VII) is A. purple B. colourless C. green D. pink

555) Consider the reaction represented by the equation below:
$CH_{4(g)} + 2O_{2(g)} -----> CO_{2(g)} + 2H_2O_{(l)}$. How many moles of carbon (IV) oxide will be produced from 32g of methane? (H = 1, C = 12, O = 16) A. 2 B. 3 C. 4 D. 8

556) Fats are classified as A. hydrocarbons B. alkanoates C. alkanols D. carbohydrates

557) Ethyne undergoes the following reactions except A. polymerization B. addition C. substitution D. esterification

558) Which of the following compounds has the highest percentage of carbon? [H = 1, C = 12, O = 16] A. C_2H_2 B. C_2H_4 C. C_2H_6 D. C_2H_6O

559) Hardness in water can be removed by adding A. copper (II) tetraoxo-sulphate (IV) B. sodium trioxocarbonate (IV) C. sodium chloride D. alum

560) How many orbitals are in the d-sub shell? A. 1 B. 3 C. 5 D. 7

561) An element X has isotopic masses of 6 and 7. If the relative abundance is 1 to 12.5 respectively, what is the relative atomic mass of X?
A. 6.0 B. 6.1 C. 6.9 D. 7.0

562) An atom $^{238}_{92}X$ decays by alpha particle emission to give an atom Y. The atomic number and mass number of Y are A. 90 and 234 respectively B. 91 and 238 respectively C. 92 and 236 D. 93 and 238 respectively

563) An element with mass number 133 and atomic number 55 has A. 55 electrons and 55 neutrons B. 55 electrons and 78 neutrons C. 78 electrons and 78 neutrons D. 78 electrons and 55 neutrons

564) Which of the following pairs of species contains the same number of electrons? [$_6C$, $_8O$, $_{10}Ne$, $_{11}Na$, $_{12}Mg$, $_{13}Al$, $_{17}Cl$] A. Mg^{2+} and Al^{3+} B. Cl^- and Ne C. Na^+ and Mg D. C and O^{2-}

565) An element X has electronic configuration $1s^2 2s^2 2p^6 3s^2 3p^6 4s^2$
To which group of the periodic table does X belong? A. I B. II C. III D. IV

566) Which of the following electronic configurations represent that of a noble gas? A. 2, 8, 8, 2 B. 2, 8, 2 C. 2, 8 D. 2, 6

567) Diamond is a hard substance because its carbon atoms are held by
A. delocalized electrons B. strong electrostatic forces C. Van der Waals forces
D. strong directional covalent bonds

568) The presence of unpaired electrons in an atom of a d-block element accounts for its A. ductility B. luster C. malleability D. paramagnetism

569) The atomic numbers of elements X and Y are 20 and 17 respectively. Which of the following compounds is likely to be formed by the combination of the two elements? A. XY B. XY_2 C. XY_3 D. X_2Y

570) What type of bond will be formed between elements P and Q if their electronegativity values are 0.8 and 4.0 respectively? A. Covalent bond B. Co-ordinate bond C. Ionic bond D. Metallic bond

571) What type of chemical bonding is involved in the formation of NH_4^+ from a molecule of ammonia and a proton? A. Covalent bonding B. Co-ordinate bonding C. Electrovalent bonding D. Hydrogen bonding

572) What is responsible for metallic bonding? A. Attraction between the delocalized electrons and fixed positive lattice points (cations) B. Attraction between positive and negative ions C. Sharing of electrons between the metal atoms D. Transfer of electrons from one atom to another

Use the following portion of the periodic table to answer Questions 16 to 18

II	II	III	IV	V	VI	VII	VIII
A							B
	C		D				
E						F	G

51

573) Which of the letters indicate elements which exist as diatomic gases?
A. B and G   B. C and F   C. C and A   D. A and F

574) Which of the letters represents an alkaline earth metal?   A. F B. E C. D D. C

575) Which of the following pairs of letters denotes elements that have the same number of electrons in their outermost shells?   A. C and D   B. E and F   C. B and G   D. A and B

576) If 20 cm^3 of distilled water is added to 80cm^3 of 0.50 mol dm^{-3} hydrochloric acid, the new concentration of the acid will be   A. 0.10 mol dm^{-3} B. 0.30 mol dm^{-3} C. 0.40 mol dm^{-3} D. 2.00 mol dm^{-3}

577) Consider the reaction represented by the equation below.
$2NaHCO_{3(s)} + Heat \longrightarrow Na_2CO_{3(s)} + CO_{2(g)} + H_2O_{(g)}$.
Which of the following can be used to test for the gas produced?
A. Caustic soda   B. Lime water   C. Bromine water   D. Quicklime

578) Which of the following gases will have the lowest rate of diffusion under the same conditions?
[N = 14, O = 16, Cl = 35.5, Ar = 40] A. Argon B. Chlorine C. Nitrogen D. Oxygen

579) Consider the reaction: $H^+_{(ag)} + OH^-_{(aq)} \longrightarrow H_2O_{(l)}$
The energy change taking place in the reaction above is enthalpy of
A. formation   B. hydration   C. neutralization   D. solution

580) Which of the following processes is an endothermic reaction? A. Dissolving NH_4Cl crystals in water B. Addition of concentrated H_2SO_4 to water   C. Dissolving NaOH pellets in water   D. Passing SO_3 gas into water

581) Which of the following methods cannot be used to distinguish between a strong acid and a weak acid?   A. Conductivity measurement   B. Measurement of pH   C. Measurement of heat of neutralization D. Action on starch iodide paper

582) The indicator used in neutralizing CH_3COOH and NaOH solutions has pH range of   A. 3 – 5
B. 7 – 8   C. 8 – 10   D. 10 – 12

583) Which of the following compounds absorbs moisture from the atmosphere and dissolves in it?
A. $FeCl_3$ B. $MgSO_4.7H_2O$ C. Na_2SO_4 D. KCl

584) Consider the reversible reaction represented by the equation below:
$2SO_{2(g)} + O_{2(g)} \longrightarrow 2SO_{3(g)}$: Enthalpy change = -197 kJ mol^{-1}

Which of the following conditions will **not** increase the yield of sulphur (VI) oxide? A. Increase in temperature B. Decrease in temperature C. Increase in pressure D. Addition of O_2 into the mixture

585) Electrolysis is applied in the following processes **except** A. electroplating B. extraction of aluminium C. extraction of iron D. purification of copper

586) The oxidation number of iodine in the iodate ion (IO_3^-) is A. -5 B. -1 C. +1 D. +5

587) The product of the complete oxidation of ethanol will be an A. alkane B. alkanal C. alkanoic acid D. alkanone

588) Which of the following industrial processes is chlorine **not** used? A. Production of polyvinylchloride (PVC) B. Manufacturing of hydrochloric acid C. Manufacturing of common salt D. Manufacturing of domestic bleach

589) Which of the following compounds is an alkanoate? A. CH_3COOH B. CH_3COOCH_3 C. CH_3CH_2OH D. CH_3CH_2COOH

590) What is C_aH_b in the following equation? $C_aH_b + 5O_2 \longrightarrow 3CO_2 + 4H_2O$
A. C_3H_4 B. C_3H_6 C. C_3H_8 D. C_5H_{10}

591) Greenhouse effect can be reduced by controlling A. water evaporation B. burning of wood and fossil fuel C. the use of aerosols D. the use of artificial fertilizers

592) Waste plastics accumulate in the soil and pollute the environment because plastic materials are A. insoluble in water B. non-biodegradable C. easily affected by heat D. inflammable

593) Which of the following substances is an ore of iron? A. Bauxite B. Cassiterite C. Haematite D. Steel

594) When sodium atom forms the ion Na^+, A. it gains one electron B. it gains one proton C. it achieves a noble gas configuration D. its atomic number increases

595) Which of the following statements about **rare** gases are **correct**? I. Their outermost shells are fully filled II. They are generally unreactive III. Their outermost shells are partially filled IV. They have lone pair of electrons in their outermost shell A. I and II only B. II and III only C. I, II and III only D. I, II, III and IV

596) How many electrons are in the ion F^-? $[^{19}_9F]$ A. 8 B. 9 C. 10 D. 19

597) Which of the following properties is characteristic of the halogens?   A. Ability to accept electrons readily   B. Ability to donate electrons readily   C. Ability to form basic oxides   D. Formation of coloured compounds

598) In which of the following atoms is the ionic radius larger than the atomic radius?   [$_{11}$Na, $_{12}$Mg, $_{13}$Al, $_{17}$Cl]   A. Aluminium B. Chlorine C. Magnesium D. Sodium

599) A transition metal would be expected to   A. form oxides with different formulae   B. have strong oxidizing ability   C. have low density   D. have an atomic structure with a complete outermost shell

600) Which of the following properties of atoms generally **increases** down a group in the periodic table? A. Electron affinity   B. Electronegativity   C. Ionic radius   D. Ionization energy

601) Which of the following statements about sodium is **not** true?   A. Na$^+$ is smaller in size than Na   B. Na$^+$ is larger in size than Na C. Na$^+$ has fewer electrons than Na.   D. The effective nuclear charge in Na$^+$ is greater than in Na.

602) If the difference between electronegativities of two elements is large, the type of bond that can be formed between them is   A. covalent B. dative C. ionic D. metallic

603) Which of the following species does **not** contain a co-ordinate bond?
A. Al$_2$Cl$_6$ B. CCl$_4$ C. H$_3$O$^+$ D. NH$^+_4$

604) Which of the following compounds has hydrogen bonds between its molecules?
A.   HF   B. HBr   C. HCl   D. HI

605) 14.8g of a salt (Z) dissolves in 250cm^3 of distilled water to give a concentration of 0.80 mol dm^{-3}. Calculate the molar mass of the salt (Z).
A. 13. 5g mol^{-1} B. 18.5g mol^{-1} C. 47.4 g mol^{-1} D. 74.0 g mol^{-1}

606) What is the IUPAC name of the compound represented by the molecular formula NaClO$_4$?
A.   Sodium tetraoxochlorate (I)   B. Sodium tetraoxochlorate (IV)   C. Sodium tetraoxochlorate (VI)   D. Sodium tetraoxochlorate (VII)

607) When air is successively passed through sodium hydroxide solution, alkaline pyrogallol and then concentrated tetraoxosulphate (VI) acid, its remaining components are   A. oxygen and water vapour   B. oxygen and nitrogen   C. carbon (IV) oxide and noble gases   D. nitrogen and noble gases

608) Which of the following gases is lighter than air? A. CO$_2$ B. SO$_2$ C. HCl D. NH$_3$

609) Which of the following statements is **true** of the molecules of a gas under ideal conditions? The molecules   A. move at random   B. undergo inelastic collisions   C. attract each other   D. occupy a large volume

610) 25 cm^3 of 0.80 mol dm^{-3} hydrochloric acid neutralized 20 cm^3 of sodium hydroxide solution. What is the concentration of sodium hydroxide in mol dm^{-3}? $NaOH_{(aq)} + HCl_{(aq)} \longrightarrow NaCl_{(aq)} + H_2O_{(l)}$ A. 0.08   B. 0.10   C. 0.80   D. 1.00

611) Which of the following compounds is a basic salt?   A. $Mg(NO_3)_2$ B. $(NH_4)_2SO_4$ C. $K_4Fe(CN)_6$ D. $Zn(OH)Cl$

612) What is the oxidation number of chromium in K_2CrO_4?   A. +1   B. +2   C. +4   D. +6

613) What product is formed at the cathode during the electrolysis of concentrated sodium chloride solution using carbon electrodes?   A. Chlorine   B. Hydrogen   C. Oxygen   D. Sodium

614) Which of the following substances is a suitable solvent for perfumes? A. Benzene   B. Ethanol   C. Turpentine   D. Water

615) The complete hydrogenation of C_6H_6 in the presence of nickel catalyst at 200°C gives   A. C_6H_8 B. C_6H_{10} C. C_6H_{12} D. C_6H_{14}

616) What is the empirical formula of a hydrocarbon containing 0.08 moles of carbon and 0.32 moles of hydrogen?   A. CH_2 B. CH_3 C. CH_4 D. C_2H_4

617) What is the product of the reaction between ethanol and excess acidified $KMnO_4$ solution?   A. $CH_2 = CH_2$ B. CH_3COOH C. $CH_3 - CH_3$ D. CH_3OCH_3

618) Glucose reduces Fehling's solution on warming to   A. copper (I) oxide   B. copper (II) oxide   C. copper (I) chloride   D. copper (II) hydroxide

619) Which of the following activities is not a source of air pollution?   A. Cigarette smoking   B. Carbon dating   C. Domestic fires   D. Coal powered stations

620) From which of the following ores is iron extracted?   I. Haematite   II. Bauxite   III. Magnetite   IV. Cassiterite.   A. I and II only   B. I and III only   C. I, II and III only   D. I, II, III and IV

621) Which of the following methods can be used to separate a mixture of two miscible liquids with different boiling points? A. Decantation   B. Distillation   C. Evaporation   D. Filtration

622) The following substances are heavy chemicals **except**   A. sodium trioxocarbonate (IV)   B. tetraoxosulphate (VI) acid   C. lead (IV) tetraethyl   D. sodium hydroxide

623) Pig-iron is brittle because it contains   A. a high percentage of carbon as impurity B. calcium trioxosilicate (IV)   C. unreacted haematite   D. undecomposed limestone

624) Which of these metals will **not** liberate hydrogen from dilute HCl?   A. Copper B. Iron C. Magnesium D. Zinc

625) What volume will 0.5g of H_2 occupy at s.t.p.?   [H = 1; 1 mole of a gas occupies 22.4 dm^3 at s.t.p.] A. 2.24 dm^3 B. 5.60 dm^3 C. 11.20dm^3 D. 44.80dm^3

626) The mass spectrometer can be used to measure mass of a   A. an atom B. an electron C. a proton D. a neutron

627) Find the number of neutrons in an atom represented by $^{45}_{21}X$ A. 21 B. 24 C. 45 D. 66

628) The valence electrons in a chloride ion are [$_{17}Cl$]   A. 2p electrons only   B. 3s and 3p electrons only C. 3p and 3d electrons only   D. 3p electrons only

629) What is the atomic number of an element whose cation Y^+ has the electronic configuration $1s^2 2S^2 2p^6$?   A. 9 B. 10 C. 11 D. 12

630) Atom $^{234}_{88}Q$ decay by alpha emission to give an atom **R.** The atomic number and mass number of atom R are respectively   A. 86 and 230   B. 87 and 234   C. 88 and 233   D. 90 and 238

631) Which of the following statements is **not** true of halogens? A. They exist in different physical states   B. They exist as diatomic molecules   C. Their ionic radii decrease down the group   D. Their melting and boiling points increase down the group

632) Elements in the same group of the periodic table have   A. the same number of valence electrons   B. the same number of electron shells   C. different chemical behaviour   D. the same atomic size

633) Water molecules are held together by   A. covalent bond   B. hydrogen bond C. ionic forces D. van der Waals forces

634) The force of attraction between covalent molecules is   A. dative bonding B. hydrogen bonding C. ionic force   D. van der Waals force

635) What is the mass of 6.02×10^{24} atoms of magnesium?   [Mg = 24, Avogadro constant = 6.02×10^{23} atoms mol^{-1}]   A. 240 g B. 24 g   C. 2.4 g D. 0.24 g

636) If 5.00cm^3 of 0.20 mol dm^{-3} Na$_2$CO$_3$ was diluted with distilled water to obtain 250 cm^3 solution, what is the concentration of the resulting solution? A. 0.004 mol dm^{-3} B. 0.02 mol dm^{-3} C. 0.20 mol dm^{-3} D. 0.40 mol dm^{-3}

637) Which of the following gases has the lowest rate of diffusion under the same conditions of temperature and pressure? [H = 1, O = 16, Ne = 20, CI = 35.5]   A. Chlorine B. Hydrogen C. Neon D. Oxygen

638) In a fixed volume of a gas, an increase in temperature results in an increase in pressure due to an increase in the   A. number of collisions between the gas molecules   B. number of repulsion between the gas molecules C. number of the collisions between the gas molecules and the walls of the container   D. kinetic energy of the gas

639) When concentrated H$_2$SO$_4$ is added to NaCl$_{(s)}$, the gas evolved   A.   bleaches damp blue litmus paper   B. forms a white precipitate with AgNO$_{3(aq)}$ C. forms a white precipitate with BaCl$_{2(aq)}$ D. turns moist red litmus paper blue

640) The collision theory proposes that   A.   reactants collide more frequently to bring about reduction in the reaction rate B.   all of the collisions of reactants are effective C.   reactants must collide with a   certain minimum frequency for a reaction to take place.   D. collision of reactants should be equal to collision of products

641) Diamond does **not** conduct electricity because it   A. has no free valence electrons B.   is a giant molecule   C. contains no bonded electrons   D. is a solid at room temperature

642) The colour of phenolphthalein indicator in dilute HNO$_{3(aq)}$ is   A.   colourless   B.   orange C. pink D. purple

643) An aqueous solution of CaCl$_2$ gives   A. acidic solution   B.   alkaline solution   C. a buffer solution D. a neutral solution

644) Which of the following compounds would dissolve in water to give a solution whose pH is less than 7?   A. Al(NO$_3$)$_3$ B. KNO$_3$ C. N$_2$O   D. NH$_3$

645) Consider the reaction represented by the equation below:
KOH$_{(aq)}$ + HCl$_{(aq)}$ -----> KCl$_{(aq)}$ + H$_2$O$_{(l)}$. What volume of 0.25 mol dm^{-3} KOH would be required to completely neutralize 40 cm^3 of 0.10 mol dm^{-3} HCl?   A. 40cm^3 B. 32 cm^3 C. 24 cm^3 D. 16 cm^3

646) When 100 cm^3 of a saturated solution of KCIO$_3$ at 40C is evaporated, 14 g of the salt is recovered. What is the solubility of KCIO$_3$ at 40°C? [KCIO$_3$ = 122.5]
A. 11.42 mol dm^{-3} B. 8.80 mol dm^{-3} C. 1.14 mol dm^{-3} D. 0.88 mol dm^{-3}

647) The high solubility of ethanol in water is due to its   A. low boiling point   B.   low   freezing point   C. covalent nature   D. hydrogen bonding

648) Calculate the quantity of electricity passed when 0.4 A flows for 1 hour 20 minutes through an electrolytic cell   A. 4800 C   B. 3840 C   C. 1920 C   D. 32 C

649) A mixture of kerosene and diesel oil can be separated by   A. crystallization   B.   distillation   C. precipitation   D. sublimation

650) Which of the following compounds would not give a precipitate with ammoniacal $AgNO_3$ solution?   A. $CH_3C≡CCH_3$ B. $HC≡CH$ C. $CH_3C≡CH$ D. $CH_3CH_2C≡CH$

651) When ethanol is heated with excess concentrated tetraoxosulphate (VI) acid, the organic product formed is   A. ethanal B. ethanoic acid C. ethane D. ethene

652) The hydrolysis of groundnut oil by potassium hydroxide is known as   A. hydrogenation B. saponification C. esterification D. neutralization

653) What is the main type of reaction alkenes undergo?   A. Addition B. Condensation C. Elimination D. substitution

654) A hydrocarbon containing 88.9% carbon has the empirical formula [H = 1 C = 12] A. CH   B. CH_2 C. C_2H_3 D. C_2H_5

655) A solution of a salt was acidified with HCl. When a few drops of $BaCl_2$ solution were added, a white precipitate was formed. Which of the following anions is present in the salt?   A. CO^{2-}_3 B. NO^-_3 C. SO^{2-}_4 D. SO^{2-}_3

656) Which of the following substances causes the depletion of the ozone layer in the atmosphere?   A. Carbon (IV) oxide   B. Chlorofluorocarbon C. Sulphur (IV) oxide D. Hydrocarbons

657) Chemical that are produced in small quantities and with very high degree of purity are   A. bulk chemicalsB. fine chemicals C. heavy chemicals D. light chemicals

658) Aluminium is a good roofing material because it   A. is strong and does not conduct heat and electricity B. readily absorbs light   C. forms an oxide film over its surface which prevents corrosion   D. is flexible and can conduct heat

659) Which of the following alloys is **mainly** used in making statues?
A. Brass   B. Bronze   C. Duralumin D. Steel

660) Which of the following scientists discovered the neutron? A.   Ernest Rutherford   B.   J. J. Thompson   C. James Chadwick D. R. A. Millikan

58

661) How many protons does $^{40}_{20}Ca$ contain?　A. 20　B. 30　C. 40　D. 60

662) If 100 atoms of element X contains 70 atoms of ^9X and 30 atoms of ^{11}X, calculate the relative atomic mass of X　A. 9.6　B. 10.6　C. 11.6　D. 20.0

663) How many orbitals are associated with the p-sub energy level? A. 2 B. 3 C. 5 D. 6

664) The following properties are characteristics of transition elements except
A. formation of complex ions　B. fixed oxidation states　C. formation of coloured compounds
D. catalytic abilities

665) An element Q forms a compound QCl_5. In which group of the periodic table is Q?
A. I　B. III　C. V　D. VII

666) Which of the following arrangements is in order of increasing ionization energy?
A. Al, Si, P, S　B. Si, Al, S, P　C. S, P, Si, Al　D. P, Si, S, Al

667) Which of the following properties of covalent compounds is **not** correct?
They　A. are non-electrolytes　B. are mostly gaseous and volatile liquids　C. have low
melting　points　D.　have　high　boiling　points

668) Which of the following molecules is not linear in shape?　A. CO_2　B. O_2　C. NH_3　D. HCl

669) In bonded atoms, increase in electronegativity difference,　A. Increases polarity　B.
decreases polarity　C. has no effect on polarity　D. brings the polarity to zero

670) The number of particles in one mole of a chemical compound is the　A. atomic number B.
Avogadro's number　C. mass number　D. oxidation number

671) Which of the following gases is lighter than air?　A. HCl B. SO_2　C CO_2　D. NH_3

672) What is is the value of -14°C on the Kelvin temperature scale? A. 259K　B. 259°K　C. 287K　D. 287°K

673) In a mixture of gases which do not react chemically together, the pressure exerted by the individual gases is called　A. atmospheric pressure　B. partial pressure　C. total pressure
D. vapour pressure

674) Which of the following dilute acids does not react with metals to liberate hydrogen?　A. HNO_3　B. H_2SO_4　C. HCl　D. CH_3COOH

675) Which of the following hydroxides will readily dissolve in water? A. $Cu(OH)_2$ B. NaOH C. $Pb(OH)_2$ D. $Zn(OH)_2$

676) Which of the following acids forms normal salt only? A. Tetraoxosulphate (VI) acid B. Tetraoxophosphate (V) acid C. Trioxonitrate (V) acid D. Trioxosulphate (IV) acid

677) A solution that contains as much solute as it can dissolve at a given temperature is said to be A. concentrated B. saturated C. supersaturated D. unsaturated

678) Which of the following salts is insoluble in water? A. $Pb(NO_3)_2$ B. Na_2CO_3 C. $AgNO_3$ D. AgCl

679) Which of the following statements is **correct**? The solubility of A. gases increases with increase in temperature B. gases decreases with increase in temperature C. most solid solutes decreases with increase in temperature D. most solid solutes is constant

680) A catalyst increases the rate of chemical reaction by A. decreasing the temperature of the reaction B. decreasing the activation energy of the reaction C. increasing the surface area of the reactants D. decreasing the surface area of the products

681) Which of the following substances is the anode in the dry Leclanche cell? A. Carbon rod B. Muslin bag C. The seal D. Zinc container

682) Consider the reaction represented by the equation: $2H_2SO_{4(aq)} + C_{(s)} \longrightarrow 2H_2O_{(l)} + 2SO_{2(g)} + CO_{2(g)}$. H_2SO_4 is acting as A. a catalyst B. an oxidizing agent C. a reducing agent D. a sulphonating agent

683) Which of the following oxides of nitrogen has oxidation number of + 1? A. KNO_3 B. N_2O C. N_2O_4 D. NO

684) The separation of petroleum fractions depends on the differences in their A. boiling points B. molar masses C. melting points D. solubilities

685) Which of the following raw materials is used in the plastic industry? A. Calcium B. Ethene C. Hydrogen D. Methane

686) When protein is heated to a high temperature it undergoes A. condensation B. denaturation C. hydrolysis D. polymerization

687) What is the molecular formula of a compound with empirical formula CH_2O and vapour density of 90? [H = 1, C = 12, O = 16] A. $C_4H_8O_2$ B. $C_4H_8O_3$ D. $C_6H_{10}C_5$ D. $C_6H_{12}O_6$

688) Ethanol reacts with excess acidified $K_2Cr_2O_7$ to produce A. ethanal B. ethane C. ethanoic acid D. ethylethanoate

689) Dry hydrogen chloride when dissolved in methylbenzene A. turns blue litmus paper red B. turns red litmus paper blue C. does not affect litmus paper D. bleaches litmus paper

690) The **major** product in the solvay process is A. NaOH B. Na_2CO_3 C. NH_3 D. H_2SO_4

691) Which of the following alloys does **not** contain copper? I. Brass II. Bronze III. Steel
A. I only B. II only C. III only D. II and III only

692) An important medical use of nuclear radiations is A. activation analysis B. carbon dating C. Radiotherapy D. tissue regeneration

693) Which of the following elements exhibits the same chemical properties as the atom $^{35}_{17}X$? An element with A. atomic number 17 B. atomic number 18 C. mass number 35 D. mass number 52

694) Which of the following noble gases has electronic structure similar to that of N in NH_3? [$^{14}_{7}N$]
A. $_2He$ B. $_{10}Ne$ C. $_{36}Ar$ D. $_{36}Kr$

695) The energy change which accompanies the addition of an electron to a gaseous atom is A. atomization B. electron affinity C. electronegativity D. ionization

696) Which of the following elements is a d-block element? A. Calcium B. Iron
C. Lithium D. Sillicon

697) Which of the following elements is diatomic? A. Iron B. Neon C. Oxygen D. Sodium

698) Calcium and magnesium belong to the same group in the periodic table because both A. are metals B. form cations C. form colourless salts D. have the same number of valence electrons

699) Which of the following elements is polyatomic? A. Iron B. Neon C. carbon D. Sulphur
700) Which of the following statements about chlorine and iodine at room temperature is correct?
A. Chlorine is gas and iodine is solid B. Chlorine is liquid and iodine is gas C. Chlorine and iodine are gases D. Chlorine is solid and iodine is liquid

701) If X is a group II element, its oxide would be represented as A. X_3O_2 B. X_2O C. X_2O_3
D. XO

702) Which of the following species correctly represents an ion of M with 13 protons and 10 electrons? A. $_{10}M^{3+}$ B. $_{10}M^3$ C. $_{13}M^{3+}$ D. $_{13}M^{3-}$

703) A solid substance with high melting and boiling points is likely to be a/an A. covalent compound B. dative covalent compound C. electrovalent compound D. non-metal

704) Which of the following molecules has a linear shape? A. CH_4 B. CO_2 C. H_2S D. NH_3

705) Which of the following molecules has a triple bond in its structure? A. CH_4 B. NH_3 C. N_2 D. O_2

706) The bonds in crystalline ammonium chloride are A. covalent and dative B. ionic and covalent C. ionic, covalent and dative D. ionic, covalent and hydrogen bond

707) A solution of sodium trioxocarbonate (IV) contains 10.6g in 250 cm^3 of solution. Calculate the concentration of the solution [Na_2CO_3 = 106.0]
A. 0.4 mol dm^{-3} B. 1.0 mol dm^{-3} C. 10.6 mol dm^{-3} D. 25.0 mol dm^{-3}

708) What is the volume occupied by 2 moles of ammonia at s.t.p? A. 44.8 dm^{-3} B. 22.4 dm^{-3} C. 11.2 dm^{-3} D. 5.6 dm^{-3}

709) Which of the following gases contains the highest number of atoms at s.t.p? A. 6 moles of neon B. 3 moles of oxygen C. 2 moles of chlorine D. 1 mole of ethane

710) Given that r is rate and p is density, the expression $r \propto 1/\sqrt{p}$ represents
A. Boyle's Law B. Charles' Law C. Dalton's law D. Graham's Law

711) The determination of heat of combustion is carried out with A. a thermometer B. a bomb calorimeter C. an evaporating dish D. a boiling tube

712) The minimum amount of energy required for effective collisions between reacting particles is known as A. activation energy B. bond energy C. kinetic energy D. potential energy

713) Which of the following oxides is basic? A. NO_2 B. Al_2O_3 C. SO_2 D. CaO

714) Which of the following equimolar solutions would have the highest conductivity? A. $NH_4NO_{3(aq)}$ B. $NaNO_{3(aq)}$ C. $Mg(NO_3)_{2(aq)}$ D. $Al(NO_3)_{3(aq)}$

715) Which of the following chlorides in insoluble in water? A. $AgCl$ B. KCl C. NH_4Cl D. $ZnCl_2$

716) Which of the following factors would not affect the solubility of a gas? A. Concentration B. Nature of solvent C. Pressure D. Temperature

717) The rate of chemical reaction of solids are **not** affected by
A. catalyst B. pressure C. particle size D. temperature

718) Which of the following statements about the cell notation $Mg/Mg^{2+}//Cu^{2+}/Cu$ is **correct**? A. Copper is the anode B. Magnesium is reduced C. Magnesium is the anode D. The double line represents the electrodes

719) Which of the following statements about the electrolysis of $CuSO_{4(aq)}$ using copper cathode and platinum anode is **not** correct?   A. Copper is deposited at the cathode B. Oxygen is liberated at the anode   C. It is used for the purification of copper D. The solution becomes acidic

720) The quantity of electricity required to discharge 1 mole of univalent ion is   A. 9,600 C.   B. 48,250 C.   C. 96,500 C.   D. 193,000 C.

721) Fats and oils are used as raw materials in the following industries **except** A.   paint industry B. plastic industry C. margarine industry D. cosmetics industry

722) Which of the following substances is trihydric? A. Ethanol B. Glycol C. Glycerol D. Phenol

723) An advantage of detergent over soap is that detergents   A. are readily available   B. are in powdered form C. are non-biodegradable D. Lather readily with water

724) The products of fermentation of sugar are   A. carbon (IV) oxide and water   B. Ethanol and carbon (IV) oxide   C. ethanol and water   D. ethanol and enzymes

725) The IUPAC name of $C_2H_5COOC_2H_5$ is   A.   ethylethanoate   B.   ethylpropanoate   C. propylethanoate D. Propylpropanoate

726) An organic compound contains 40.0% carbon, 6.7% hydrogen and 53.3 oxygen. What is the empirical formula of the compound? [O = 16.0,   C = 12.0, H = 1.0]   A. C_2HO B. CHO   C. CH_2O D. CHO_2

727) Compound **N** reacts with sodium metal to produce a gas that gives a 'pop' sound with a burning splint, **N** also reacts with ethanoic acid to give a sweet smelling liquid.   Compound **N** is an   A. Alkanol B. alkanoate C. alkane D. alkanoic acid

728) The **main** function of limestone in the blast furnace is to   A. act as a reducing agent   B. act as a catalyst C. remove impurities   D. supply oxygen

729) Which of the following metals exists as liquid at ordinary temperature?   A. Copper B. Gold C. Mercury D. Silver

730) How many isotopes has hydrogen? A. 2 B. 3 C. 4 D. 5

731) What type of reaction is represented by the following equation?
$^2_1D + ^3_1H \longrightarrow ^4_2He + ^1_0n + energy$ A. Nuclear fission   B. Nuclear fusion
C. Radioactive decay D. Spontaneous decay

732) Which of the following ions has the electron configuration 2, 8, 8?   A. Na^+ B. Mg^{2+} C. F^- D. Cl^-

733) An element with the electron configuration of $1s^2$ $2s^2$ $2p^6$ would have a combining power of
      A. 0 B. 2 C. 6 D. 8

734) Rare gases are stable because they   A. contain equal number of protons and neutrons   B. contain more electrons than protons C. are chemically active D. have octet structure

735) Which of the following elements would produce coloured ions in aqueous solution?   A. Calcium B. Zinc C. Aluminium D. Manganese

736) Which of the following hydrohalic acids is the weakest?   A. HBr B. HCl C. HF D. HI

737) Chlorine, bromine and iodine belong to the same group and   A. are gaseous at room temperature B. form white precipitate with $AgNO_{3(aq)}$ C. react violently with hydrogen without heating   D. react with alkali

738) Which of the following elements can conveniently be placed in two groups in the periodic table?   A. Carbon B. Copper   C. Hydrogen D. Oxygen

739) The bond formed when two electrons that are shared between two atoms are donated by only one of the atoms is   A. covalent B. dative C. ionic D. metallic

740) When element $_{20}A$ combines with element $_8Y$,   A. a convalent compound, AY is formed   B. an ionic compound, AY is formed   C. an ionic compound, A_2Y is formed   D. a covalent compound, AY_2 is formed

741) In metallic solids, the forces of attraction are between the mobile valence electrons and   A. atoms   B. neutrons C. the negative ions   D. positively charged nuclei

742) Which of the following statements about displacement reaction is **correct**?   A. A more electropositive element displaces a less electropositive one   B. A less electropositive element displaces a more electropositive one   C. The position of elements in the reactivity series has no effect on the reaction   D. It only occurs when the reaction is at equilibrium

743) The volume occupied by 17g of H_2S at s.t.p is   [H = 1.00, S = 32.0, Molar volume = 22.4 dm^3]
      A. 11.2 dm^3 B. 17.0 dm^3 C. 34.0 dm^3 D. 44.8 dm^3

744) What is the amount of magnesium that would contain 1.20×10^{24} particles? [Mg = 24, Avogadro's constant = 6.02×10^{23}]   A. 0.5 moles B. 2.0 moles C. 12.0 moles D. 24.0 moles

745) The number of atoms in one mole of a substance is equal to the   A.   mass   number   B. oxidation number C. atomic number D. Avogadro number

746) Which of the following statements about a molar solution is **correct?** It A. is a supersaturated solution B. cannot dissolve more of the solute at that temperature C. contains any amount of solute in a given volume of solution D. contains one mole of the solute in 1 dm^3 of solution

747) A gas that is collected by upward delivery is likely to be A. heavier than air B. insoluble in water C. lighter than air   D. soluble in water

748) Bubbling excess carbon (IV) oxide into calcium hydroxide solution results in the formation of   A. $CaCO_3$ B. CaO C. $Ca(HCO_3)_2$ D. H_2CO_3

749) The equation P = K/V illustrates   A. Boyle's law B. Charles' law C. Dalton's law D. Gay Lussac's law

750) The initial volume of a gas at 300 K was $220cm^3$. Determine its temperature if the volume became 250 cm^3 A. 183 K   B. 264 K   C. 300 K   D. 341 K

751) Which of the following statements about enthalpy of neutralization is **correct?** It   A is constant for a strong acid and a strong base   B. cannot be determined using calorimeter   C. has a positive value   D. is higher for strong acid and a weak base

752) When $NH_4Cl_{(s)}$ was dissolved in water, the container was cold to touch. This implies that   A. the process is endothermic B. the process is exothermic   C. NH_4Cl is highly soluble in water D. NH_4Cl forms a saturated solution

753) Which of the following metallic oxies is amphoteric?   A. Al_2O_3 B. Fe_2O_3 C. MgO D. Na_2O

754) On evaporation to dryness, 250 cm^3 of saturated solution of salt **X** with relative molar mass 101 gave 50.5g of the salt. What is the solubility of the salt? A. 1.0 mol dm^{-3} B. 2.0 mol dm^{-3} C. 4.0 mol dm^{-3} D. 5.0 mol dm^{-3}

755) Consider the following reversible reaction equation: $X_{(g)} + Y_{(g)} \text{-----> } XY_{(g)}$ Enthalpy change = -220 KJ mol^{-1}.   If the temperature of the system is increased, the A. backward reaction would be favoured   B. forward reaction would be favoured   C. reaction would stop D. reaction would be at equilibrium

756) Which of the following conditions would lead to an increase in the rate of a reaction? A. Increase in temperature and decrease in the surface area of reactants B. Increase in both temperature and concentration of reactants C. Decrease in temperature and increase in concentration of reactants D. Decrease in temperature and increase in the surface area of reactants

757) What mass of copper would be formed when a current of 10.0 A is passed through a solution of $CuSO_4$ for 1 hour? [Cu =63.5; 1F = 96500 C] A. 5.9g B. 11.8g C. 23.8g D. 37.3g

758) Which of the following metals could be used as sacrificial anode for preventing the corrosion of iron? A. Copper B. Lead C. Magnesium D. Silver

759) Which of the following compounds determines the octane rating in petrol? A. 1, 2, 3 – trimethypentane B. 2, 3, 5 – trimethyloctane C. 2, 3, 5 – trimethylpentane D. 2, 2, 4 – trimethylpentane

760) Which of the following compounds would react with ethanoic acid to give a sweet smelling liquid? A. Alkane B. Alkanol C. Alkanal D. Alkyne

761) Which of the following separation techniques would show that black ink is a mixture of chemical compounds? A. Crystallization B. Chromatography C. Filtration D. Sublimation

762) The following substances are examples of addition polymer **except** A. nylon, B. Perspex C. polyethane D. polychloroethane

763) When bromine is added to ethane at room temperature, the compound formed is A. 1, 1 – dibromoethane B. 1, 1 – dibromoethene C. 1, 2 – dibromoethane D. 1, 2 – dibromoethene

764) The compound that makes palm wine taste sour after exposure to the air for few days is A. ethanol B. ethanoic acid C. methanol D. methanoic acid

765) The reagent that can be used to distinguish ethane from ethyne is A. ammoniacal silver trioxonitrave (V) solution B. Benedict solution C. bromine water D. Fehling's solution

766) The following substances are ores of metals **except** A. bauxite B. cuprite C. cassiterite D. graphite

767) Which of the following processes does **not** involve the use of limestone? A. Extraction of iron in the blast furnace B. Manufacture of tetroxosulphate (VI) acid by contact process C. Production of washing soda by Solvay process D. Production of cement

768) Which of the following substances is **mainly** responsible for the depletion of the ozone layer? A. chlorofluorocarbon B. Carbon (IV) oxide C. Nitrogen D. Oxygen

769) Aluminium is extracted by electrolysis from A. bauxite B. cryolite C. duralumin D. kaolin

770) The number of orbitals in a f-sub level of an atom is A. 2 B. 3 C. 5 D. 7

771) The atomic radii of metals are usually A. greater than their ionic radii B. equal to their ionic radii C. less than their ionic radii D. less than those of non-metals in the same period

772) Two elements X and Y are in the same group on the period table because they both have the same A. number of electronic shells B. number of valence electrons C. atomic size D. atomic number

773) The d-block elements are paramagnetic because they A. contain paired electrons which are repelled by magnetic filed B. contain unpaired electrons in the partially filled 3d-orbital C. form coloured complex ions that attract magnetic lines of force D. have delocalized valence electrons

774) Which of the following statements about a radioactive substance is/are **correct**? I. it emits radiation continuously and spontaneously II. The emitted radiations are affected by temperature and pressure III. The radiation can penetrate opaque matter A. II only B. I and II only C. I and III only D. II and III only

775) Which of the following elements is a metalloid? A. Carbon B. Oxygen C. Silicon D. Sodium

776) Which of the following halogens is liquid at room temperature? A. Chlorine B. Fluorine C. Iodine D. Bromine

777) The shape of a graphite crystal is A. tetrahedral B. pyramidal C. hexagonal D. octahedral

778) The compound formed by the combination of two elements with a large electronegativity difference is likely to be A. polar covalent B. giant molecular C. covalent D. ionic

779) The complex compound formed when aluminium dissolves in aqueous sodium hydroxide is A. $Na_2Al(OH)_4$ B. $NaAl(OH)_4$ C. $NaAl(OH)_3$ D. $Na_2Al(OH)_3$

780) MgO does not readily dissolve in water because A. of its high melting point B. it is a covalent compound C. it forms a hydroxide when dissolved in water D. its lattice energy is higher than its hydration energy

781) The formula of mercury (I) dioxonitrate (III) is   A. $HgNO_3$ B. Hg_2NO_2 C. $Hg_2(NO_2)_2$ D. $Hg(NO_3)_2$

782) A sample of a gas may be identified as chlorine if it turns   A. damp blue litmus paper red B. lime water milky C. lead ethanoate paper black D. starch iodide paper blue-black

783) A metal M forms two types of chlorides, **MCl_2** and **MCl_3** which of the following laws **best** explains the relationship between the chlorides? Law of
A. conservation of mass B. definite proportion C. multiple proportion   D. reciprocal proportion

784) Which of the following metals would readily displace hydrogen from steam? A. Copper B. Lead C. Magnesium D. Silver

785) The volume occupied by 0.4 g of hydrogen gas at s.t.p is
(H = 1.00; Molar volume at s.t.p = 22.4 dm^3)   A. 4.48dm^3 B. 11.2dm^3 C. 0.896dm^3 D. 8.96dm^3

786) When a substance changes directly from the gaseous state to the solid state without forming a liquid, the substance is said to   A. condense B. evaporate C. sublime D. precipitate

787) At ordinary temperature H_2O is a liquid while H_2S is a gas. This is because H_2O has A. weak intermolecular forces holding its molecules together   B. strong hydrogen bonds holding its molecules together   C. induced dipole-dipole forces between its molecules D. ionic forces between its molecules

788) The postulate that molecules are in constant random motion **best** explains why liquids   A. can undergo solidification   B. maintain their volumes C. are incompressible D. have no characteristic shape

789) Which of the following gases has the **lowest** rate of diffusion under the same condition? [H = 1.00; He = 4.00; O = 16.0; Cl = 35.5]
A. Cl_2 B. H_2 C. He D. O_2

790) The energy evolved when magnesium burns in air is in the form of   A. heat B. heat and sound C. light and heat D. sound

791) A substance L reacts with $NH_4NO_{3(aq)}$ to generate ammonia gas. L is likely to be   A. HCl B. NaOH C. CH_3COOH D. $CaSO_4$

792) Which of the following pH values indicates that a solution is a strong base?
A. 1 B. 5 C. 9 D. 13

793) The hydrolysis of NH$_4$Cl gives   A. an acidic solution   B. an alkaline solution   C. a buffer solution   D. a neutral solution

794) A spot of oil paint on a shirt can **best** be removed using   A. brine   B. detergent   C. kerosene D. warm water

795) Consider the reaction represented by the following equation :
CuO$_{(s)}$ + H$_2$SO$_{4(aq)}$ -----> CuSO$_{4(aq)}$ + H$_2$O$_{(l)}$

Which of the following factors will **not** affect the rate of the reaction?
A. Particle size of CuO$_{(s)}$ B. Concentration of H$_2$SO$_{4aq}$ C. Temperature of the reacting mixture
D. Pressure reaction system

796) Which of the following devices function on redox reaction?
I. Dry cell   II. Car battery   III. Electric generator
A. I and III only   B. II and III only   C. I and II only   D. I, II and III

797) Consider the following reaction equation   C$_{16}$H$_{34}$ -----> C$_5$H$_{12}$ + C$_{11}$H$_{22}$
The process represented by the equation is A. cracking   B. fermentation   C. polymerization   D. reforming

798) Which of the following substances would **not** produce ethanol when fermented?   A. Cane sugar   B. Glucose   C. Starch   D. Vinegar

799) An alkanol can be prepared by the reaction of an alkene with   A. concentrated tetraoxosulphate (VI) acid   B. bromine in tetrachloroethane   C. aqueous potassium tetraoxomanganate (VII)   D. sodium hydroxide solution

800) A compound contains 7.75% hydrogen, 38.21% carbon and 54.04% chlorine. Determine the empirical formula of the compound [H = 1.00; C = 12.0; Cl = 35.5] A. C$_2$H$_3$Cl   B. C$_2$H$_5$Cl   C. C$_3$H$_8$Cl   D. C$_5$H$_2$Cl

801) Which of the following industrial processes depends on the action of enzymes?   A. Liquefaction of air   B. Manufacture of soap   C. Brewing of beer   D. Catalytic cracking

802) Which of the following pollutants is **not** usually recycled?   A. Aluminium cans   B. Glass bottles   C. Nuclear wastes   D. Paper wastes

803) A metal that is **widely** used in the manufacture of paints and overhead electric cables is   A. aluminium   B. copper   C. iron   D. lead

804) Brass is a mixture of   A. Cu and Sn   B. Cu and Zn   C. Cu and Mg   D. Cu and Pb

805) Which of the following substances is **mainly** responsible for the depletion of the ozone layer?
A. Oxygen B. Chlorofluorocarbon C. Carbon (II) oxide D. Nitrogen (II) oxide

806) Which of the following instruments is used in detecting the presence of radiation?   A. Cathode ray tube B. Geiger-Muller counter C. Mass spectrometer D. X-ray tube

807) The molecule which has a linear shape is   A. CH_4 B. NH_3 C. H_2S D. CO_2

808) The formula of the compound formed between a trivalent metal, **M** and a divalent non-metal, **Y** is   A. M_2Y_3 B. M_3Y_2 C. MY D. M_3Y

809) An atom of an element X gains two electrons. The symbol of the ion formed is A. X^+, B. X^{2+} C. X^{2-} D. X

810) Which of the following statements is correct?   A. Atomic size decreases down the group B. Atomic size increases across the period   C. Anions are smaller than the parent atom   D. Cations are smaller than the parent atom

811) The element with electron configuration $1s^2 2s^2 2p^6 3s^2 3p^1$ belongs to   A. s-block, period 3, group 1   B. p-block, period 3, group 2   C. s-block, period 3, group 3   D. p-block, period 3, group 3

812) In the periodic table, all the elements within the same group have the same   A. number of neutrons B. number of valence electrons C. number of isotopes D. atomic number

813) Which of the following halogens is liquid at room temperature?   A. Iodine   B. Chlorine C. Bromine D. Fluorine

814) Rare gases are stable because they A. are chemically active   B. contain equal number of protons and neutrons C. contain more electrons than protons D. have octet structures

815) In the periodic table, alkaline earth metals can be found in group A. I B. II C. VI D. VII

816) Which of the following bond types is responsible for the high boiling point of water? A. Metallic bond   B. Covalent bond C. Ionic bond D. Hydrogen bond

817) In metallic solid, the forces of attraction is between the mobile valence electrons and the   A. atoms B. neutrons C. negative ions D. positively charged nuclei

818) The bonds in crystalline ammonium chloride are   A. covalent and dative   B. ionic and covalent C. ionic, covalent and dative D. ionic, covalent and hydrogen bond

819) Which of the following elements is diatomic? A. Sodium B. Oxygen C. iron D. Neon

820) Noble gas molecules are held together by   A. van der Waals forces B. hydrogen bonds C. dative bonds D. covalent bonds

821) Which of the following statements about nuclear reaction is **correct?** The reaction   A. involves neutrons only B. takes place inside the nucleus C. is governed by temperature and pressure D. involves protons and electrons only

822) Consider the reaction represented by the following equation:
$C_2H_2 + yH_2 \longrightarrow C_2H_6$. The value of y in the reaction is   A. 4 B. 3 C. 2 D. 1

823) The volume of 0.25 mol dm^{-3} solution of KOH that would yield 6.5g of solid KOH on evaporation is (K − 39.0; O = 16.0; H = 1.00) A. 464.30 cm^3 B. 625.00 cm^3 C. 1000.00 cm^3 D. 2153.80 cm^3

824) The gas law which describes the relationship between volume and temperature is A. Boyle's law B. Charles' law C. Dalton's law D. Graham's law

825) Which of the following phenomena leads to decrease in volume of a liquid in an open container?   A. Brownian motion B. Diffusion C. Evaporation D. Sublimation

826) The pressure exerted by a gas is a function of the   A. total volume of the gas B. speed of the gaseous molecules C. mass of each gaseous molecules D. frequency of collision between gaseous molecules

827) Which of the following variables is a measure of the average kinetic energy of the molecule of a gas? A. Density B. Pressure C. Temperature D. Volume

828) When heat is absorbed during a chemical reaction, the reaction is said to be   A. adiabatic B. endothermic C. exothermic D. isothermal

829) The aqueous solution which has pH > 7 is   A. $FeCl_{3(aq)}$ B. $CuSO_{4(aq)}$ C. $KNO_{3(aq)}$ D. $Na_2CO_{3(aq)}$

830) Which of the following compounds crystallizes without water of crystallization? A. $MgSO_4$ B. Na_2CO_3 C. NaCl D. $FeSO_4$

831) A substance is said to be **impure** if   A. its melting point range is wide B. it dissolves in water with difficulty C. it has a low melting point D. it is coloured

832) The following factors affect the solubility of a solid in a given solvent **except**   A. nature of solute B. nature of solvent C. pressure D. temperature

833) Consider the reversible reaction represented by the equation:

$N_2O_{4(g)}$ -----> $2NO_{2(g)}$: $\triangle H = + x \, KJ \, mol^{-1}$. An increase in pressure will
A. favour the forward reaction B. increase the yield of NO_2 C. increase the yield of N_2O_4
D. have no effect on the reaction

834) Which of the following cells produce electrical energy from chemical reactions?
I. Daniel Cell II. Solar Cell III. Lead Acid Accumulator IV. Generator
A. I and II only B. I, II and III only C. I and III only D. I, III and IV only

835) What happens at the cathode during electrolysis? The   A. anion is oxidized B. anion loses
electrons C. cation is oxidized D. cation is discharged

836) Which of the following substances are electrolytes? I. $PbBr_{2(l)}$ II. $NaCl_{(aq)}$ III. $NaCl_{(s)}$ IV.
$C_6H_{12}O_{6(aq)}$ A. I only B. I and II only C. II only D. III and IV only

837) The separation of petroleum fractions depends on the differences in their
A. melting points B. molar masses C. solubilities D. boiling points

838) The **major** product formed by the reaction between ethanoic acid and aqueous sodium
hydroxide is   A. soap B. sodium ethanoate C. sodium methoxide D. water

839) Which of the following organic compounds would decolourize bromine water?
A. Benzene B. Cyclobutane C. Hexane D. Pentane

840) How many isomers has $C_2H_4Cl_2$?   A. 2 B. 3 C. 4 D. 5

841) Which of the following reactions is common to all hydrocarbons?
A. Combustion   B. Addition   C. Polymerization D. Condensation

842) A hydrocarbon compound contains 92.3% carbon. Determine its empirical formula   [H =
1.00;   C = 12.0]   A. CH. B. CH_2 C. CH_3 D. C_2H_3

843) The main function of limestone in the blast furnace is to   A. act as catalyst
B. act as reducing agent   C. remove impurity D. supply carbon (IV) oxide

844) Which of the following raw materials is used in a plastic industry?
A. Ethene B. Methane C. Calcium D. Hydrogen

845) Which of the following statements about thermoplastic material is **correct?** They
A. do not melt on heating B. harden on heating C. decompose on heating D. soften and melt
on heating

846) Bronze is a mixture of   A. Cu and Mg B. Cu and Sn C. Cu and Zn D. Cu and Pb

847) Which of the following statements about fine chemical is **correct?** It   A. is injurious to health   B. has low degree of purity   C. is produced in relatively small amount   D. can be stored for a long time

848) Which of the following materials is classified as a non-biodegradable pollulant? A. Animal hide   B. Paper   C. Plastic   D. Wood

849) Which of the following atoms contains complete electrons in the outermost shell? A. $_8O$ B. $_{10}Ne$ C. $_{15}P$ D. $_{19}K$

850) Le Chatelia's Principle is related to   A. quantum numbers of electrons   B. reversibility of equilibrium reactions   C. electronegativity values of elements   D. collision theory of reaction rates

851) Which of the orbitals 4s, 2p, 3d and 4f has the lowest number of electrons?   A. 4s   B. 2p   C. 3d   D. 4f

852) Which of the following has the highest charge?   A. Alpha particle   B. Deuterium   C. Gamma ray   D. Neutron

853) Four radioactive elements P, Q, R and S have half-life periods of 10 hours, 20 hours, 30 hours and 40 hours respectively. Which of them is most stable?   A. P   B. Q   C. R   D. S

854) Zinc atom ionizes by   A. gaining two electrons   B. losing two electrons   C. sharing two electrons   D. gaining two protons

855) Which of the following properties decreases across a period in the Periodic Table?   A. Atomic radius   B. Electronegativity   C. Electron affinity   D. Ionization energy

856) In the Periodic Table, the elements that gain electrons most readily belong to   A. Group I   B. Group II   C. Group III   D. Group VII

857) Which of the following halogens is a solid at room temperature?   A. F_2 B. Br_2 C. I_2 D. Cl_2

858) The molecular bonds in water are held together by   A. electrostatic forces   B. van der Waals' forces   C. hydrogen bonds   D. covalent bonds

859) Exhaust fumes discharged from a smoky vehicle gradually fades away as a result of   A. diffusion   B. osmosis   C. absorption   D. emission

860) Which of the following is not a direct product of the destructive distillation of coal?   A. Soot   B. Coal tar   C. Coke   D. Coal gas

861) Equal volumes of SO_2 cnd SO_3 at s.t.p. have the same A. mass B. density C. rate of diffusion D. Avogadro's number

862) Which of the following does not fully exhibit the characteristics of transition metals? A. Manganese B. Chromium C. Zinc D. Iron

863) The contact process is used for the manufacture of A. Na_2CO_3 B. H_2SO_4 C. HCl D. NH_3

864) Which of the following gases is lighter than air? A. SO_2 B. CO_2 C. NH_3 D. NO_2

865) Which of the following element exhibits allotropy? A. Sulphur B. Hydrogen C. Nitrogen D. Chlorine

866) An endothermic reaction is one which A. is very hot B. gives out heat C. gives out smoke D. absorbs heat

867) Which of the following gives an alkaline solution? A. NH_4Cl B. $MgSO_4 . 7H_2O$ C. CH_3COONa D. $Zn(NO_3)_2$

868) Which of the following compounds dissolves in water to form a solution with pH equal to 7? A. Sodium tetraoxosulphate (VI) B. Potassium hydroxide C. Ammonium chloride D. Sodium trioxocarbonate (IV)

869) If the solubility of $CaCO_3$ is 0.20 mol dm^{-3} at room temperature, calculate the mass of $CaCO_3$ in 200cm^3 of the solution at this temperature. [$CaCO_3$=100g mol^{-1}] A. 4.0g B. 10.0g C. 40.0g D. 100.0g

870) The property of tetraoxosulphate (VI) acid which makes it useful as a drying agent is that it is A. ionic B. hygroscopic C. a strong electrolyte D. a strong acid

871) Which of the following lowers the activation energy of a chemical reaction? A. Freezing mixture B. Reducing agent C. Water D. Catalyst

872) Which of the following gas is monatomic? A. Oxygen B. Nitrogen C. Hydrogen D. Neon

873) Which of the following is the best electrolyte?. A. Glucose B. Sodium trioxonitrate (V) C. Ethanoic acid D. Silver chloride

874) The following compounds may be regarded as diacidic except: A Potassium hydroxide B. Calcium hydroxide C. Zinc hydroxide D. Magnesium hydroxide

875) Separating funnel is used for separating a mixture of A. water and sand B. gold and sand C. water and kerosene D. petrol and ethanol

876) How many hydrogen atoms are present in one molecule of cyclobutane?   A. 6   B. 8   C. 10   D. 12

877) Which of the following is the most reactive towards chlorine?   A. Methane   B. Ethene   C. Ethyne   D. Ethanol

878) Which of the following compounds is saturated?   A. 2-methylbut-1-ene   B. cyclopropane   C. Methylbenzene   D. ethyne

879) Which of the following has no isomer?   A. $C_2H_2Cl_2$ B. C_4H_9Cl C. C_4H_{10} D. C_3H_8

880) Which of the following is the major impurity during the extraction of iron in the blast furnace?   A. Sulphur   B. Limestone   C. Silicon (IV) oxide   D. Carbon

881) The following are non-biodegradable except   A. hair   B. finger nails   C. plastics   D. Hides

882) Which of the following is an amphoteric oxide?   (a) ZnO   (b) CO_2 (c) SO_2 (d) Na_2O

883) A gas that is collected by upward delivery is likely to be   A. heavier than air   B. insoluble in water   C. lighter than air   D. soluble in water

884) Which of the following elements is a d-block element?   A. Calcium   B. Iron   C. Lithium   D. Silicon

885) Which of the following chlorides is insoluble in water?   A. AgCl   B. KCl   C. NH_4Cl D. NaCl

886) Consider the equilibrium reversible reaction represented by the following equation: $2SO_{2(g)}$ + $O_{2(g)}$ → $2SO_{3(g)}$ ΔH = -395.7KJ/mol. Which of the following statements about the equilibrium system is correct?   A. Addition of catalyst changes the equilibrium position   B. Decrease in pressure increases the yield of SO_3 C. Decrease in pressure increases the equilibrium concentration of O_2 D. Increase in temperature favours the forward reaction

887) The rate of chemical reaction of solids are not affected by: A. catalyst   B. pressure   C. particle size   D. temperature

888) Which of the following salts will dissolve in water to give a pH value greater than 7?   A. $FeCl_3$ B. $CuSO_4$ C. KNO_3 D. Na_2CO_3

889) Which of the following organic compounds would decolourize bromine water?   A. Benzene   B. Cyclobutane   C. Hexane   D. Pentane

890) Which of the following elements would produce coloured ions in aqueous solution? A. Ca B. Fe C. Mg D. Na

891) All the following trioxocarbonate(iv) are insoluble in water except (a) Na_2CO_3 (b) $CuCO_3$ (c) $PbCO_3$ (d) $CaCO_3$.

892) Which pollutant gas is produced by the decomposition of vegetation? A carbon (II) oxide B methane C nitrogen (IV) oxide D sulphur (IV) oxide

893) Which type of compound is characterized by the presence of –OH group? A alcohol B alkane C alkene D carboxylic acid

894) X, Y and Z are three hydrocarbons. X CH2=CH2 Y CH3–CH=CH2 Z CH3–CH2–CH=CH2. What do compounds X, Y and Z have in common? 1 They are all alkenes. 2 They are all part of the same homologous series. 3 They all have the same boiling point. A 1, 2 and 3 B 1 and 2 only C 1 and 3 only D 2 and 3 only

895) Which statements about ethanol are correct? 1 It can be made by fermentation. 2 It is an unsaturated compound. 3 It burns in air and can be used as a fuel. A 1, 2 and 3 B 1 and 2 only C 1 and 3 only D 2 and 3 only

896) Which of the following is not a direct product from crude oil? (a) gasoline (b) ethanol (c) kerosene (d) bitumen

897) What is the shape of a molecule of H_2O? A. Pyramidal B. Tetrahedral C. Trigonal planar D. Non-linear

898) Which of the following increases as solid melts? A. Temperature of the solid B. Degree of disorder of the solid molecules C. Number of molecules D. Activation energy

899) Which of the following gases shows a brown ring when present in its test? A. NO_2 B. SO_2 C. NH_3 D. CO_2

900) The gas evolved when dilute tetraoxosulphate (VI) acid reacts with common salt is A. hydrogen chloride B. oxygen C. carbon (IV) oxide D. sulphur (IV) oxide

901) If a solution has a pH of 11, it can be concluded that it A. is a weak electrolyte B. has hydroxide ion concentration of 0.11 mol dm^{-3} C. is twice as alkaline as a solution of pH 5.5. D. will have more OH^- than H^+

902) Which of the following compounds dissolves in water to give an alkaline solution? A. Ammonium chloride B. Sodium chloride C. Sulphur (IV) oxide D. Potassium trioxocarbonate (IV)

903) Which of the following salts will give a coloured solution? A. K_2SO_3 B. $(NH_4)_2$ SO_4 C. $NaHCO_3$ D. $AgNO_3$

904) What type of reaction is involved when coal is heated to produce coke?   A.   Redox reaction   B. Destructive distillation   C. Dehydration   D. Neutralization

905) Which of the following metals is the most reactive? A. Aluminium   B. Magnesium   C. Iron   D. Zinc

906) In which of the following is the oxidation number of hydrogen equal to -1?
A. H_2 B. H_2S C. NaH   D. H_2O

907) 0.10 mol dm^{-3} $MgSO_4$ conducts electricity better than 0.10 mol dm^{-3} CH_3COOH because the solution of $MgSO_4$ A. contains more ions   B. is neutral to litmus   C. has a lower molar mass   D. has a higher pH.

908) C_2H_2 will undergo the following reactions except   A.   cracking   B.   combustion   C. substitution   D. addition

909) The following pairs belong to the same class except   A. glucose and sucrose   B. maltose and lactose   C. sucrose and maltose   D. fructose and galactose

910) In the electrolytic extraction of sodium from fused sodium chloride, molten calcium chloride is added in order to   A. lower the melting point of the sodium chloride   B. prevent over heating of the sodium chloride   C. lower the activation energy of the reaction   D. form a protective crust on top of the electrolyte

911) Which of the following requires the process of salting out?   A. baking process   B. wine production C. production of antibiotics   D. manufacture of soap

912) Which of the following is not a polymer? A. Starch B. Protein   C. Fat D. PVC

913) Allotropy is exhibited by the following elements except: A. Chlorine B. carbon C. Oxygen D. sulphur

914) What is the mass number of an element if its atom contains 15 protons, 18 electrons and 16 neutrons?   A. 31   B. 33   C. 15   D. 34

915) W, X, Y and Z are elements in the same period in the Periodic Table but in groups 3, 4, 5 and 6 respectively. Which of them gains electrons most readily?
A. W   B. X   C. Y   D. Z

916) The atom of an element Y has one electron in its outermost shell. What is the formula of the compound formed when Y combines with aluminium ($_{13}Al$)?   A. AlY_2 B. Al_3Y C. Al_2Y_3 D. AlY_3

917) Which of the following compounds is electrovalent? A. $CaCl_2$ B. CO C. H_2O D. CH_4

918) An aluminium sheet is placed in the path of a beam from a radioactive source. The emissions that pass through the sheet consist of   A. alpha particles and gamma rays   B. alpha and beta particles   C. gamma rays only   D. beta particles only

919) Which of the following features of a bone sample can be determined by carbon-dating? A. Thickness   B. Mass   C. Age   D. Animal source

920) Which of the following pairs of reagents react to produce an organic compound? A. Zinc and concentrated trioxonitrate (V) acid   B. Water and calcium carbide   C. Copper an dilute hydrochloric acid   D. Magnesium and dilute tetraoxosulphate (VI) acid

921) A mixture of potassium chloride and zinc trioxocarbonate (IV) in water can be separated by   A. evaporation   B. sublimation   C. distillation   D. Filtration

922) A piece of limestone was dropped into a beaker containing dilute HCl. The gas evolved was   A. NH_3 B. SO_2 C. NO_2 D. CO_2

923) From the reaction in the question above, it can be concluded that limestone   A. contains salt   B. is acidic   C. contains a trioxocarbonate (IV)   D. has a pH greater of 7.

924) What is the oxidation number of phosphorus in $Ca_3(PO_4)_2$?   A. +1   B. +3   C. +5   D. +6

925) Which of the following substances will not conduct electric current?   A. Glucose solution   B. Zinc rod   C. Hydrochloric acid   D. a solution of common salt in water

926) A concentrated solution containing H^+, Na^+, OH^- and Cl^- was electrolysed using platinum electrodes. Which of the substances below will be discharged at the anode?   A. Hydrogen   B. Sodium   C. Oxygen   D. Chlorine

927) Partial hydrogenation of propyne yields   A. butyne   B. propane   C. propene   D. propanol

928) Ethyne produces more luminous flame than ethene on burning because ethyne   A. has a higher molar mass   B. has a higher proportion of carbon   C. undergoes both substitution and addition reactions   D. is a gas at room temperature

929) Ethanol burns in air to produce   A. ethanal   B. ethanoic acid   C. Oxygen   D. carbon (IV) oxide

930) Which of the following is a polyamide?　A. Nylon　B. Polyvinyl chloride　C. Polyethene D. Cellulose

931) Which of the following industries uses ester in its operation?　A. Fertilizer plant B. Textile industry　　C. Brewery　D. Soap manufacturing industry

932) Which of the following is not an effect in the use of hard water?　A. Formation of scum with soap　B. Furring of kettles　C. discolouration of water pipes　D. Formation of strong bones

933) Which of the following compounds is readily soluble in water?　A. CuO B. AgCl　C. $MgSO_4$ D. $CaCO_3$

934) If $10cm^3$ of distilled water is added to $20cm^3$ of dilute HCl, the concentration of the acid　A. increases　B. decreases　C. remains constant　D. doubles

935) Which of the following is not an electrolyte　A. lime water　B. brine　C. acidified water　D. ethanol

936) The atomic number of chlorine is 17. What is the number of electrons in a chloride ion? A. 16　　B. 17　C. 18　D. 19

937) An element has the electronic configuration, $1s^2 2s^2 2p^6 3s^2 3p^1$. To what group does it belong in the periodic table?　A. 1　B. 2　C. 3　D. 6

938) Which of the following gases can be produced by using Kipp's apparatus? A. H_2　B. Cl_2　C. H_2S　D. NH_3

939) Which of the following acids can form two acid salts?　A. Tetraoxosulphate (VI) acid　B. Trioxosulphate (IV) acid　C. Tetraoxophosphate (V) acid D. Trioxonitrate (V) acid

940) A substance is said to be efflorescent if it loses　A. water of crystallization to the atmosphere B. heat to the surrounding　C. carbon (IV) oxide to the atmosphere　D. moisture to the atmosphere and changes colour

941) Which of the following substances changes to the gaseous state when heated in air?　A. Sodium chloride B. Iodine crystals　C. Magnesium ribbon　D. Copper (II) oxide

942) Which of the following salt solutions will have a pH equal to 7?
A. $Na_2SO_{3(aq)}$　B. $Na_2CO_{3(aq)}$　C. $Na_2SO_{4(aq)}$　D. $NaHSO_{4(aq)}$

943) Which of the following compounds will not release a gas when heated?　　A. $Cu(NO_3)_2$ B. $AgNO_3$　C. K_2CO_3　D. $CaCO_3$

944) Which of the following will give a colourless solution?   A. $FeSO_4$ B. $Fe_2(SO_4)_3$ C. $CuSO_4$ D. $MgSO_4$

945) When brine is electrolyzed using inert electrodes the products are   A. oxygen and hydrogen   B. hydrogen and chlorine   C. sodium and oxygen   D. sodium and chlorine

946) Which of the following processes takes place during the alkaline hydrolysis of vegetable oils?   A. Esterification   B. Hydrolysis   C. Hydrogenation   D. Saponification

947) Which of the following compounds can be represented by the molecular formula $C_2H_4O_2$?   A. Ethanal   B. Ethanol   C. Ethanoic acid   D. Ethanone

948) When alkynes are hydrogenated completely, they produce compounds which can undergo the following reactions except   A. substitution B. addition   C. combustion   D. cracking

949) The following compounds contain double bond except   A. methylpropanoate   B. propanoic acid   C. Benzene   D. cyclopropane

950) Which of the following will decolourise bromine water?   A. cyclopentane   B. Ethanoic acid   C. propyne   D. propanol

951) The reaction of lactose with dilute HCl produces   A. alkanoate   B. glucose and fructose   C. a black mass of carbon   D. glucose and galactose

952) Which of the following molecules is not linear in shape?   A. CO_2 B. O_2 C. N_2 D. H_2O

953) Which of the following gases is an alkaline gas?   A. HCl   B. CO_2 CO_2 D. NH_3

954) What is the value of $-50°C$ on the Kelvin temperature scale?   A. $223°K$ B. $223K$ C. $287K$ D. $287°K$

955) In a mixture of gases which do not react chemically together, the pressure exerted by the individual gases is called   A. atmospheric pressure   B. partial pressure   C. total pressure   D. vapour pressure

956) Which of the following salts is insoluble in water?   A. $Pb(NO_3)_2$ B. Na_2CO_3 C. $AgNO_3$ D. $CaSO_4$

957) Which of the following substances is the anode in the dry Leclanche cell?   A. Carbon rod   B. Muslin bag   C. The seal   D. Zinc container

958) Which of the following oxides of nitrogen has oxidation number of +1? A. NO_2 B. N_2O C. N_2O_4 D. NO

959) When protein is heated to a high temperature it undergoes A. condensation B denaturation C. hydrolysis D.polymerization

960) Which of the following elements reacts violently with water? A. Iron B. Aluminium C. Magnesium D. Potassium

961) Which of the scientist put forward the atomic theory? A. Dalton B. Maxwell C. Chadwick D. Rutherford

962) Group 0 element are: A. polyatomic B. triatomic C. diatomic D. monoatomic

963) The order of elements in the periodic table is according to A. mass number B. proton number C. valency number D. number of neutrons

964) An oxide is represented as XO_2. Which of the following non-metals cannot be X? A. Carbon B. Chlorine C. Nitrogen D. Sulphur

965) In which of the following compounds is hydrogen bond absent? A. HF B. H_2O C. CH_3CH_2OH D. NaH

966) The movement of water molecules from a region of higher concentration to a region of lower concentration is A. Diffusion B. Osmosis C. Brownian movement D. Kinetic movement

967) Which of the following compounds can be used to increase the pH of a solution from 2 to 7? A. SO_2 B. HCl C. KCl D. CaO

968) The salt Na_2HPO_4 is a/an A. normal salt B. basic Salt C. acid salt D. Double salt

969) Which of the following oxide is acidic? A. CO B. N_2O C. MgO D. SO_2

970) When benzene is completely hydrogenated, it produces A. hezane B. cyclohexane C. hexene D. cyclohexene

971) A reaction between propene and iodine is regarded as A. substitution B. Addition C. polymerization D. Elimination

972) Which of the following compounds has the molecular formula C_2H_6O? A. Ethanal B. Ethanol C. Ethanone D. Ethanamide

973) What is the IUPAC name of $HCOOC_3H_7$? A. Methylethanoate B. Methylpropanoate C. Propylmethanoate D. Propylethanoate

974) Galvanized iron is iron plated with A. Copper B. Tin C. Zinc D. Silver

975) Why is sodium stored in paraffin oil? A. It is an alkali metal B. It is very reactive with water C. It is very reactive with air D. It forms a basic oxide

976) Which of the following catalyst is used in the contact process? A. Pellets of iron B. Platinium C. Vanadium (V) oxide D. Nickel

977) Which of the following elements is not present in stainless steel? A. Carbon B. Iron C. Chromium D. Copper

978) Which of the following is a mixed anhydride? A. CO_2 B. NO_2 C. SO_2 D. SO_3

979) Which of the following is the formula for sand? A. SiO_2 B. P_2O_5 C. BeO D. ZnO

980) Which of the following elements is a liquid at room temperature? A. Lithium B. Krypton C. Mercury D. Barium

981) Which of the following molecules is linear in shape? A. CO_2 B. N_2O C. NH_3 D. Na_2S

982) A weak base A. is not concentrated B. ionized in water to produce OH^- C. is also an alkali D. has no effect on litmus.

983) A solution that forms white precipitate with $BaCl_2$ contains A. NO_3^- B Ca^{2+} C. SO_3^{2-} D. Cl^-

984) Consider the reaction: $Cl_2 + HI \longrightarrow HCl + I_2$. The reducing agent is A. Cl_2 B. HI C. HCl D. I_2

985) What mass of iron will be deposited by the liberation of Fe^{2+} when 0.1F of electricity flows through an aqueous solution of a iron (II) tetraoxosulphate (VI)? [Fe = 56] A. 56g B. 28g C. 5.6g D. 2.8g

986) Fats and oils are classified as A. alkanoic acids B. Organic bases C. esters D. alkanols

987) $H_3O^+_{(aq)} + OH^-_{(aq)} \longrightarrow 2H_2O_{(l)}$. Which of the following can produce the H_3O^+? A. CaO B. HCl C. CO D. NH_3

988) Which of the following hydrocarbons is saturated? A. Cyclohexane B. Benzene C. But-1-ene D. Ethyne

989) The solid product of the destructive distillation of coal is A. Coke B. coal tar C. ammoniacal liquor D. coal gas.

990) Which of the following processes involve hydrolysis?   A. Fermentation B. Formation of amino acid from protein C. Polymerization of ethyne D. Esterification

991) Which of the following element is present in duralumin?   A. Zinc B. Lead C. Iron D. Manganese

992) Which of the following can be found in the uncombined state in nature?   A. Iron B. Magnesium C. Silver D. Aluminium

993) Which of the following oxides is basic?   A. Fe_2O_3 B. CO_2 D. NO E. SnO_2

994) Which of the following compounds is hygroscopic? A. NaOH   B. $MgSO_4.7H_2O$ C. H_2SO_4 D. $CaCl_2$

995) The oxidation number of phosphorus in PO_4^{3-} is   A. +6   B. +1   C. +7 D. +5

996) Alkanols undergo oxidation to produce the following except   A. alkanoate B. alkanal C. alkanoic acid   D. alkanone

997) Which of the following has charge but no mass? A. Alpha particle B. Neutron   C. Beta particle D. Proton

998) Which of the following compounds is an acid?   A. HCOOH B. CH_3COOCH_3 C. CH_3CH_2O D. CH_3CHO

999) Greenhouse effect gas does not include A. CO_2 B. CH_4 C. CFC   D. He

1000) Which of the following substances is an ore of Tin?   A. Bauxite B. Cassiterite C. Haematite D. Steel

ANSWERS

1. D	21. E	41. A	61. A	81. E	101. B
2. B	22. A	42. C	62. C	82. D	102. B
3. E	23. D	43. E	63. D	83. D	103. C
4. E	24. C	44. C	64. C	84. B	104. B
5. B	25. E	45. B	65. A	85. C	105. A
6. A	26. C	46. D	66. E	86. D	106. B
7. D	27. C	47. A	67. E	87. D	107. C
8. C	28. E	48. E	68. B	88. C	108. C
9. B	29. C	49. D	69. C	89. A	109. D
10 .C	30. E	50. C	70. D	90. B	110. D
11. C	31. E	51. A	71. C	91. D	111. D
12. C	32. E	52. E	72. A	92. C	112. E
13. C	33. D	53. A	73. B	93. A	113. D
14. D	34. B	54. B	74. C	94. E	114. E
15. E	35. E	55. E	75. C	95. C	115. C
16. C	36. A	56. B	76. E	96. C	116. C
17. E	37. B	57. D	77. C	97. B	117. D
18. E	38. A	58. E	78. E	98. B	118. D
19. D	39. D	59. E	79. C	99. E	119. E
20. E	40. C	60. B	80. B	100. E	120. D

121. C	141. C	161. A	181. D	201. C	221. E
122. C	142. B	162. A	182. A	202. A	222. E
123. B	143. D	163. B	183. A	203. D	223. A
124. C	144. A	164. C	184. A	204. D	224. B
125. C	145. A	165. C	185. C	205. B	225. D
126. E	146. D	166. A	186. D	206. D	226. C
127. A	147. D	167. D	187. B	207. A	227. C
128. B	148. D	168. B	188. E	208. A	228. E
129. B	149. C	169. A	189. D	209. A	229. E
130. B	150. E	170. B	190. D	210. B	230. D
131. B	151. D	171. D	191. E	211. A	231. A
132. D	152. D	172. E	192. B	212. E	232. A
133. D	153. C	173. B	193. E	213. C	233. B
134. A	154. C	174. D	194. C	214. A	234. B
135. C	155. B	175. D	195. A	215. B	235. C
136. C	156. C	176. A	196. E	216. B	236. E
137. B	157. C	177. B	197. B	217. B	237. C
138. E	158. D	178. C	198. C	218. D	238. C
139. D	159. C	179. D	199. C	219. B	239. D
140. A	160. D	180. D	200. C	220. D	240. A

241. E	261. E	281.A	301. E	321.D	341. A
242. A	262. A	282.A	302. A	322.D	342. C
243. B	263. C	283.C	303. D	323.C	343. D
244. A	264. C	284.D	304. E	324.E	344. A
245. C	265. B	285.A	305. D	325.C	345. B
246. C	266. C	286.A	306. D	326.A	346. A
247. C	267. D	287.E	307. C	327.C	347. B
248. A	268. D	288.B	308. E	328.E	348. B
249. B	269. C	289.C	309. D	329. A	349. D
250. B	270. A	290.D	310. E	330. B	350. C
251. D	271. B	291.C	311. A	331. B	351. A
252. D	272. A	292.D	312. A	332. A	352. B
253. B	273. A	293.C	313. E	333. D	353. D
254. D	274. E	294.C	314. C	334. C	354. D
255. D	275. E	295.E	315. B	335. D	355. B
256. A	276. A	296.E	316. C	336. B	356. C
257. D	277. D	297.B	317. A	337. A	357. D
258. A	278. C	298.D	318. D	338. A	358. C
259. C	279. B	299.C	319. B	339. A	359. C
260. E	280. E	300.C	320. D	340. B	360. C

361. A	381. A	401. A	421. B	441. D	461. B
362. B	382. C	402. A	422. B	442. D	462. C
363. A	383. B	403. B	423. D	443. D	463. C
364. A	384. C	404. D	424. A	444. C	464. D
365. C	385. B	405. A	425. C	445. D	465. A
366. D	386. A	406. B	426. C	446. B	466. D
367. C	387. C	407. D	427. D	447. B	467. C
368. A	388. D	408. C	428. C	448. D	468. A
369. C	389. C	409. C	429. B	449. B	469. B
370. C	390. C	410. D	430. A	450. D	470. B
371. B	391. A	411. D	431. A	451. A	471. B
372. D	392. D	412. D	432. B	452. D	472. D
373. B	393. C	413. C	433. D	453. B	473. B
374. B	394. A	414. D	434. B	454. C	474. D
375. D	395. B	415. C	435. C	455. B	475. B
376. D	396. B	416. C	436. A	456. B	476. A
377. C	397. D	417. C	437. D	457. A	477. A
378. D	398. A	418. B	438. A	458. D	478. D
379. D	399. A	419. B	439. A	459. D	479. A
380. A	400. B	420. C	440. D	460. C	480. C

481. A	501. C	521. C	541. C	561. C	581. D
482. A	502. D	522. A	542. C	562. A	582. C
483. B	503. A	523. C	543. A	563. B	583. A
484. A	504. C	524. D	544. C	564. A	584. A
485. B	505. A	525. A	545. B	565. B	585. C
486. C	506. D	526. B	546. B	566. C	586. D
487. B	507. D	527. A	547. C	567. D	587. C
488. C	508. C	528. B	548. B	568. D	588. C
489. D	509. C	529. C	549. B	569. B	589. B
490. C	510. C	530. C	550. B	570. C	590. C
491. B	511. C	531. C	551. C	571. B	591. B
492. C	512. C	532. B	552. C	572. A	592. B
493. D	513. C	533. B	553. A	573. D	593. C
494. D	514. B	534. B	554. B	574. D	594. C
495. B	515. B	535. B	555. A	575. C	595. A
496. A	516. C	536. A	556. B	576. C	596. C
497. D	517. D	537. C	557. D	577. B	597. A
498. C	518. D	538. D	558. A	578. A	598. B
499. C	519. D	539. D	559. B	579. C	599. A
500. B	520. B	540. D	560. C	580. A	600. C

601. B	621. B	641. A	661. A	681. D	701. D
602. C	622. C	642. A	662. A	682. B	702. C
603. B	623. A	643. A	663. B	683. B	703. C
604. A	624. A	644. A	664. B	684. A	704. B
605. D	625. B	645. D	665. C	685. B	705. C
606. D	626. A	646. C	666. A	686. B	706. C
607. D	627. B	647. D	667. D	687. D	707. A
608. D	628. B	648. C	668. C	688. C	708. A
609. A	629. C	649. B	669. A	689. C	709. A
610. D	630. A	650. A	670. B	690. B	710. D
611. D	631. C	651. D	671. D	691. C	711. B
612. D	632. A	652. B	672. A	692. C	712. A
613. B	633. B	653. A	673. B	693. A	713. D
614. C	634. A	654. C	674. A	694. B	714. B
615. C	635. A	655. C	675. B	695. B	715. A
616. C	636. A	656. B	676. C	696. B	716. A
617. B	637. B	657. B	677. B	697. C	717. B
618. A	638. C	658. C	678. D	698. D	718. C
619. B	639. B	659. B	679. B	699. D	719. C
620. A	640. C	660. C	680. B	700. A	720. C

721. B	741. D	761. B	781. C	801. C	821. B
722. C	742. A	762. A	782. D	802. C	822. C
723. D	743. A	763. C	783. C	803. A	823. A
724. B	744. B	764. B	784. C	804. B	824. B
725. B	745. D	765. C	785. A	805. B	825. C
726. C	746. D	766. D	786. C	806. B	826. D
727. A	747. C	767. B	787. B	807. D	827. C
728. C	748. C	768. A	788. D	808. A	828. B
729. C	749. A	769. A	789. B	809. C	829. D
730. B	750. D	770. D	790. C	810. D	830. C
731. B	751. A	771. A	791. B	811. D	831. A
732. D	752. A	772. B	792. D	812. B	832. C
733. A	753. A	773. B	793. A	813. C	833. C
734. D	754. B	774. C	794. C	814. D	834. C
735. D	755. A	775. C	795. D	815. B	835. D
736. C	756. B	776. D	796. C	816. D	836. C
737. D	757. B	777. C	797. A	817. D	837. D
738. C	758. D	778. D	798. D	818. C	838. B
739. B	759. D	779. B	799. A	819. B	839. A
740. B	760. B	780. D	800. B	820. A	840. A

841. A	861. D	881. D	901. D	921. D	941. B
842. A	862. C	882. A	902. D	922. D	942. C
843. C	863. B	883. C	903. D	923. C	943. C
844. A	864. C	884. B	904. B	924. C	944. D
845. D	865. A	885. A	905. B	925. A	945. B
846. B	866. B	886. C	906. C	926. D	946. D
847. C	867. C	887. B	907. A	927. C	947. C
848. C	868. A	888. D	908. A	928. B	948. B
849. B	869. A	889. A	909. A	929. D	949. D
850. B	870. B	890. B	910. A	930. A	950. C
851. A	871. D	891. A	911. D	931. D	951. D
852. A	872. D	892. B	912. C	932. C	952. D
853. D	873. B	893. A	913. A	933. C	953. D
854. B	874. A	894. B	914. A	934. B	954. B
855. A	875. C	895. C	915. D	935. D	955. B
856. D	876. B	896. B	916. D	936. C	956. D
857. C	877. C	897. D	917. A	937. C	957. D
858. C	878. B	898. B	918. C	938. C	958. B
859. A	879. D	899. A	919. C	939. C	959. B
860. A	880. C	900. A	920. B	940. A	960. D

961. A	981. A
962. D	982. B
963. B	983. C
964. B	984. B
965. D	985. D
966. B	986. C
967. D	987. B
968. C	988. A
969. D	989. A
970. B	990. B
971. B	991. D
972. B	992. C
973. C	993. A
974. C	994. C
975. C	995. D
976. C	996. A
977. D	997. C
978. B	998. A
979. A	999. D
980. C	1000. B

Made in the USA
Las Vegas, NV
30 January 2025